RAPTURE *in the*
MIDDLE EAST

THE MEMOIRS OF FRANCES METCALFE

The Authorized Edition by
JAMES MALONEY

WESTBOW
PRESS
A DIVISION OF THOMAS NELSON

WestBow Press books may be ordered through booksellers or by contacting:

WestBow Press
A Division of Thomas Nelson
1663 Liberty Drive
Bloomington, IN 47403
www.westbowpress.com
1-(866) 928-1240

ISBN: 978-1-4497-7116-4 (sc)
ISBN: 978-1-4497-7115-7 (hc)
ISBN: 978-1-4497-7117-1 (e)

Library of Congress Control Number: 2012919220

Printed in the United States of America

WestBow Press rev. date: 10/25/2012

CONTENTS

FOREWORD

When compiling the collected writings of the Golden Candlestick company, we had been given a mimeographed copy of Frances Metcalfe's "First Missionary Journey," her journal of an overseas trip to the Middle East in 1961, that had apparently been duplicated on a Ditto machine and preserved by Wade and Kristeen Bandelin, pastors of Shiloh Christian Ministries in Idyllwild, California. Presumably the original text is no longer extant, but the copy was readable, making transcription laborious yet fairly accurate. The journal was to be included in the three volumes of *Ladies of Gold*, but we soon realized the length of the memoirs was such that it would require its own volume. At about 135,000 words, it's the single largest writing left by Frances.

These memoirs are different than the Golden Candlestick's teachings. They're much more personal and meandering, at times, but that is the nature of one's diary: a private recollection of events that were important in one's life. I put this here because I don't want the reader to assume this is a homiletic sermon, but rather a very intimate glimpse into the life of Frances Metcalfe, a person who should be remembered and honored by the Body of Christ as a forerunner of the "seer" anointing. As such, while there are great nuggets of truth to be gleaned from this journal, there are also the everyday struggles associated with overseas travel that Frances committed to her memoirs. She talks about her hair, she talks about lost luggage, she talks about annoying tour colleagues. All of this is woven around a much deeper theme of praise and worship, perseverance, and indeed, rapture during her times in the Middle East. So you must forgive her if she seems to dwell a little too long on bad food and cold showers.

Frances was by no means perfect, I've said as much before in the other forewords. That's not disrespectful—she was a lot better off than me and just about anyone I know. But one must take into consideration that I don't unequivocally agree with every bit of philosophy and theology she puts down in this journal. I will say I agree with probably 99% of it. But I think she was perhaps a little too strict toward her traveling colleagues' humanness, and many of the "testings" she endured were most likely of her own design (I believe she would wholeheartedly agree to that statement, were she alive today) or, rather, simply the plight of living in the natural world—whereas she appears to equate a lot of them to explicit pressings from the Lord, as if she were constantly on a test. Perhaps she's right and I'm wrong. Or perhaps sometimes, they just lose your luggage.

Now that's not to say I don't believe that we as Christians are tried and tested, pressed and compressed, refined in fire. Anyone breathing on this planet can attest to such, and the Bible shows ample evidence of His people being tried in the "fiery trials" of our faith. But I do think, at least sometimes, the inconveniences and difficulties of our lives are a by-product of being in a corrupt world system, or of our own designs, in not implementing wisdom and common sense when faced with adversity.

I also want to make comment regarding Frances speaking about the Muslim god Allah as being the same as our God, Jehovah. This is an incorrect notion along the lines of universalism. Like so many Christians of her era, and even many today, she was unenlightened concerning the origins of Islam, and wrongly reasoned "Allah" just meant "God, the Covenant-Keeping One" in Arabic, simply another name for Jehovah. So under this fallacious notion, the only reason why a Muslim (or a Jew, for that matter) would not go to heaven is because they rejected God's Son, Jesus; but in essence, all three religions (Islam, Judaism and Christianity) were praying to the same Being. This is not correct, but Frances didn't know any better.

Now that we have greater access to research materials and archeological findings, and we are able to do in-depth studies concerning the origins of Islam, I want to make it clear I do not agree that Allah and Jehovah are interchangeable names for the same God. They are entirely different.

Further, concerning Judaism, while the God of the Old Testament is the same God as the New Testament, no man may come unto the Father but through Jesus Christ (John 14:6), who is very God Himself, coequal with the Father and the Spirit, triune in one God.

Still, none of her idiosyncrasies or uninformed theology devalue the good read and insights contained herein, nor should you let them bother you overly much. Frances was an amazing woman of God, and it shines through in her diary. Like all of us, she had strengths and weaknesses, but in her case, I believe the strengths vastly outweighed the weaknesses.

Some of the writing is "old-fashioned" and in some cases, might be considered callous by today's standards. She uses words like "colored folk" that we avoid today, but I assure you it was simply the vernacular of the time and did not taint Mrs. Metcalfe's love for all human beings, which is shown throughout her memoirs. I have been faithful, as in the *Ladies of Gold* volumes, to reproduce these memoirs just as she wrote them, so some archaic spelling or grammatical errors remain, but I do not think they detract from the work.

Lastly, I want to thank my daughter-in-law, Christy, for all her help in transcribing these memoirs—it was an exceptionally tedious process for all involved, and she really helped bring this project together in a timely manner.

Friends, I hope you enjoy these memoirs, and I hope you will glean not only insight into the person that was Frances Metcalfe, but also some of the biblical truths that made up the worldview of this fascinating forerunner, which in turn draw you closer to the King, our Bridegroom.

—James Maloney
Argyle, Texas
September 1, 2012

OUR FIRST MISSIONARY JOURNEY

ITINERARY

Departure: Los Angeles, May 10, 1961.

Approximate return: L.A., June 4, 5.

May 12-14	Jordan
May 15-17	Egypt
May 18-24	Israel
May 25, 26	Greece
May 27-29	Italy
May 30-June 1	France
June 2, 3	England

Letters written to Frances *must* be via AIRMAIL and it must be mailed as follows [*personal details omitted—*ed.] to reach her! *Be sure to attach sufficient postage!* In the lower left-hand corner write: c/o Pentecostal Tour. No mail will reach her mailed after the 27th. And any further correspondence should be mailed to the Mount to await her arrival home. SHALOM!

Jerusalem, Jordan
May 15, 1961

Darling ones,

We are now ready to emplane for Cairo. There is a short delay, and I am hastening to write—there hasn't been a free moment for about 48 hours. Yesterday, Mother's Day, was wonderful—first we went to the summit of the Mount of Olives—marvelous view!; the Rock of the Ascension—and a panorama of all Israel in plain view (so near and yet so far); then Gethsemane—beautiful! On to Bethlehem by these very narrow circuitous roads. So like the Bible all the way—sheep, grain, the people, the houses. All like we have pictured it (and more so!) We saw Israel again and the dividing outposts—the old road to Bethlehem. It was wonderful all over the city. After our visit at the Church and the Grotto, I bought a Bethlehem headdress. The Arab woman who sold it put it on me and I felt an anointing—Mary, Mother, Bethlehem, and all that it means. She taught me to say, "*Ana Bethlehem Sitt*" (which I believe is "I am a lady of Bethlehem"—Marian.) and a few other words. I also got a dress. I felt we were to have them. The Christian Approach Secretary met me there and our tour guides arranged to drop us at the Mission coming back. I felt led to go in wearing the headdress. This broke all the ice, and the children laughed and giggled and bowed. When I spoke to them in Arabic they were delighted. Awni is dear—but very shy. His older sister speaks English. I was overwhelmed with love for the children. And they came in and sang. They seemed starved for mother-love. I learned later that their Mission Mother has left, and they seem so eager for an American mother. They were most warm and responsive. I sang for them also—and then we all sang together. You might know, "we" adopted *two* new daughters! One is to be admitted shortly. They have 1000 or more waiting. The new one I chose—by God's help—said her name was "Noel." She is adorable, about eight. I had to choose between four—all dear. They were all so eager to be chosen that it hurt. One older girl just "adopted" me and wouldn't budge. We got many pictures.

4

These Jordanians are so lovable and responsive. The bellboys and waiters all helped teach me Arabic. They kept saying that I did very well. I spoke to everyone on the streets—as they do here. It is an amazing and incredible place.

Mike—the Secretary at the C.A.M. said the water truck saved the Mission from closing and made many friends. This year they have plenty of water. He got us a taxi back to Jerusalem and we got here in time for lunch and the afternoon tour—what roads! What driving! The hotel is fine, food delicious—and everyone just as nice as one could desire. (I did a little shopping along the way, as they give a few minutes on each tour for this.) I am just too full to even try to say it on paper. We went on the Via Dolorosa yesterday afternoon. First the Church of the Flagellation, then on. It was most impressive. Our last stop was the Garden tomb, and it was the high spot of the trip—much rejoicing and praise. Brother Mattar gives a real message. So far it has been very sad to me about the lack of praise—so much foolish talk and jesting. Most of the people are Assemblies—and I am pained, since they are ministers and teachers. They seem very ignorant about the land, and rather crude in their talk. Only the colored people seem to honor the Lord. I praise out loud—and am free—and some respond. I feel very aware of each of the Company. I keep seeing them in Arab faces—it is so dear. I feel you by turns, and again all at once. I hope this reaches you in time for Pentecost. Almost time to fly.

—Later: We are at the Jerusalem airport waiting for our plane. We had to have our bags ready by seven this morning—so it has been a long day so far. We have been sightseeing and saying our goodbyes to our charming guides and others. Last night we went to the Mount of Olives chapel. It is beautifully located and the building spacious. The Mattars are very dear and most gracious. I sang and felt anointed to witness. He is badly needing building advice, and has had some difficulties, so the dedication was postponed. How he wished Dwight could stay and supervise the work. I have been amazed at the way the Lord has opened doors for testimony, etc. Even on the street we had quite a moving with two priests of the Eastern Church. One is coming to L.A. and wanted to talk further. At Bethany it was so dear too! The exercises are rugged, but glorious.

—7 p.m. Cairo: HOT! We arrived about four and took two hours to clear customs. We are on the 15th floor of hotel "Everest." It is *noisy* but most interesting.

Accommodations are not very good here. Not much to eat today, and little sleep—I am sort of dazed, but feeling well, praise God! (Half of our party were held up in N.Y. and again in Beirut—bad weather. Praise God we made out on time. We return on jets too, have had five planes so far.

Love and kisses to all, Frances

On plane from Cairo to Jerusalem

May 19, 1961

Darlings,

It is now 8:30 a.m. and we are flying over the Sinai Peninsula. It is a most magnificent wilderness. We had a grand view of the Red Sea, Suez Canal and lower Egypt as we departed, and the clouds parted. We were having breakfast at 5:30 this morning. We were just served cakes and a sandwich. (A charming Lebanese Lt. in the army bade me a charming goodbye.) The Lord gave Acts 15:1-4 in the Amplified. And I was feeling borne on Eagle's wings, and many portions about Egypt, a letter was dropped into my lap. It was from Isabelle. It had pyramids and Scriptures about Egypt. This is the only mail received here. It was amazing to receive it while flying. I know others wrote but no mail was given out. Things were most odd in Egypt. It was dirty in our hotel (third class), and no hot water except early in the morning. Several got sick from the food. Not a soul really *understood* English in the hotel. My Arabic was priceless. I only wished I knew it better. I also used French a lot.

The people beg all the time on the streets. But some were charming, and so surprised when I would speak to them in Arabic. All were eager to help me learn. I met a darling Syrian girl by the Nile on our first night. She welcomed me in English to "*En Neel*," the river of love and peace. She was so dear and walked with us a ways. I was *forced* into Arabic, and had a wonderful time. Dwight said I was the "hit" of Cairo. I felt Mary so much, and Jesus, looking upon the unbelievable poverty and filth and at their souls with compassion.

I have had two falls, one in Beirut and one at Memphis, where I landed on my spine from a donkey cart. Only by the Lord's goodness and grace could I go on, but I'm better today. Also lost my glasses. We are in a strange mix-up. There were 590 of us in Cairo at once, and everything was odd. Sukarno was here. (The president of Indonesia.) We saw parades, and walked by the Nile Hilton—most beautiful. We saw one mosque made all of alabaster. Am so thankful to be *really* on the way to Israel. We passed thru Jordan again and thru the Mandelbaum gate. Have averaged only four hours of sleep, so I am very tired, but having the time of my life.

Much love, F.

PS. Am still sticking close to our colored friends. There is almost *no praise* or honoring of the Lord. It is surely *sad*. They are all just *tourists*. I guess I was sent to PRAISE! We are His praise in the earth. Am having liberty in witnessing.

PPS. There were almost *no* white people on the streets in Egypt. We were the "sideshow" everywhere. In the amazing Cairo Tower I met some lovely Italians. They "blessed" our going to Rome.

—This is an overwhelming experience. And our study of the nations was such a preparation.

Kings Hotel, Jerusalem, Israel
May 23, 1961

My love and kisses to all my beloved ones. It is now 7:30 a.m. and once again we are packed up and ready to leave—reluctantly—and continue to tour. I would love to remain here for at least a week. The rush, rush, rush has been most exhausting. There is no time to rest—or to write. I average about five hours in bed out of 24, and it is beginning to be felt. But the Lord is good, and gives grace. Enough has happened here to write a book about, so I will just give you a few facts—must wait for inspiration to tell much of it. I am better—my fall caused quite a lot of pain, but

I am doing much better. I got some reading glasses here, mine were not found. And my cold is going away. Some have been very ill indeed. Several have lost luggage, money or passports. There is much confusion and it takes constant vigilance to keep collected. The first person to greet us in Israel—after a rather quick crossing of "No-Man's land" was Brother Kopp. He has a son working in the Gospel here. He may build and wanted Dwight to help him also.

—Sabbath arrived Friday, and all is closed Saturday. Then Sunday was Pentecost, a very holy day here. So—I misplaced my list of those I sent cards to, so may overlook some. I am really very tired in the natural—but fine in the Spirit. Our hotel is lovely. We took Dr. McKay and Emil, his "son" who is ministering here in personal work, to dinner, $4.00 apiece. Dr. McKay took us to fillet mignon at King David Hotel. It was great after the strange food in Egypt. Our room has a view of Mount Zion. Yesterday, along with the Jews, Gypsies, Nuns, Priests and people of every kind and color, Dwight and I ascended Mount Zion for the second time. I took my harp and was mobbed by the children. I sang and played in Hebrew and English. There must have been 100 people or more who gathered. But had to leave them and get to the Upper Room. No one can go without permission—which is readily given. Five nuns were sitting on the side steps softly singing—most heavenly. I had a beautiful time on my knees. I sang softly and recited, or chanted, most of Acts 1 and 2.

—The nuns were Protestant, the Sisterhood of Mary. They sang a Pentecost song. I felt all of you!

Love, F.

"And the dove returned to him in the evening; and lo, in her mouth was a freshly plucked olive leaf."

Beloved Mary Doves,

It was with thanksgiving and rejoicing that this weary little dove made safe landing again in our nest on the Blessed Mount. The wings of the Great Dove safely bore us along through 14 takeoffs and landings, and a series of flights that took us into nine countries, in addition to our own vast nation. In two of these we visited only the airports. In seven we walked and marveled, sang and praised, and embraced the people. In the Holy Land we truly lost our heart, and more than ever before we realize that it is "the sacred heart of the world." We flew directly over many other nations, and felt a sense of beaming down the Love of Christ upon them. All in all it was an amazing *Hegira*—and was accomplished in an incredibly short time. Best of all, you flew with me, and communed with the Lord and me in every place. There was no sense of separation at all, and your fellowship in the Spirit was even more real and precious than your letters. But these were indeed welcome, and ministered to me in many ways, often supplying just the right word or thought for the very day in which I read them. Thank you and bless you each and all!

My letters to the Mount, and my cards to you, were written under difficulties. Our schedule was so tight that we were kept rushing from our meals to the bus or plane, and back again. Sleep was short and not very restful, and after we had packed or unpacked, bathed (when possible), washed our clothes and prepared for the next move there was no time for writing. Some of the cards may not have air-mail stamps, though we tried hard to obtain such. And in some cases, where we left cards and money for stamps with hotel employees, they may never have been sent at all. In any case, I did write to each of you. And I am glad Marian shared my letters with you, though they were hasty and fragmentary.

Though I was in a constantly outward and active life, the interior life of the Spirit was far more real and vivid to me than the thrilling exterior sights and experiences. I thanked God every day, and many times a day, for our years of intense discipline and training, enabling us to have abounding grace for every sort of trial, and to walk in the Spirit, no matter what the attractions or distractions around us might be. Apart from this, I would have been defeated from the start. I have never lived through a month of such constant and varied pressure and testing. Your prayers helped to keep me "singing and soaring." I am truly grateful.

It was indeed thrilling to be in beautiful Paris on the very day our President arrived, and to share with him the excitement and thrill of the lavish decorations and demonstrations in the streets. Never did Old Glory look as magnificent as when we suddenly beheld her flying high above the fountain of living waters in *The Place de la Concorde*, facing out *Avenue des Champs Elysées*. Someone said the flag was about 40 feet by 12. It was most vivid in hue and of fine silk. It waved and rippled continuously, and seemed to be fairly dancing with joy. We all sang, "The Star-Spangled Banner," and got out of the bus with much excitement. I also sang the chorus of *The Marseillaise* in honor of the French Tricolor flying just opposite. In England we equally rejoiced in billboards announcing its Gospel campaigns of Billy Graham, which are drawing large crowds. We rejoiced too in the good reports we heard of how the Pentecostal outpourings are spreading in the British Isles. How we love England and longed to visit longer there!

Our last long jaunt from London to New York, 14 hours flying time, through Passports and Customs, a long wait for another plane, and a very rough and long flight across the U.S. proved very wearing. We had to make an unexpected stop at Albuquerque for refueling. How I wanted to get off and stay off—and take a bus the rest of the way, stopping off to see Elda. But the Lord led otherwise, so back on it we piled, and found the last three hours even more trying. Something was wrong with the plane, and it made eerie noises. Los Angeles was covered with smog and the sun shining on it made it dazzling. We had no visibility of the ground until we were almost upon it. Once we headed straight toward a high mountain and seemed barely to zoom over it at the last moment. Our landing was almost like a miracle, for by this time we all felt that the pilot was lost. And in that mass of air traffic, with planes taking off and landing about one per minute, it was a little hair-raising. How good it was of the Lord to have given me the 27th chapter of Acts early in the day, and on it I stood during the entire flight. Sure enough, we all came safely to land!

Speaking of Acts 27, I am impressed that our coming gathering will center about the Apostle Paul and the account of his experiences and words in Acts, from the eighth chapter onward. At Caesarea I was strongly moved upon about this, and again I feel a witness. (You will find Caesarea mentioned again and again.) Do read all this book, and meditate upon it, as a preparation for our Anniversary meeting. It will take place June 24 and 25, D.V. beginning at eight in the evening as usual. In the meantime we have been trying to rest up and regain our lost sleep.

I am beginning to go over my notes, hoping to recapture as much as possible before the next "rush" of the Spirit and events. Thank God for years of memory training! Without it, I would have lost much that I saw and experienced, so rapidly did the scenes change. I know you are all eager for an account of the trip. And this, I am sure, can best be made in writing.

Do remember Blanche in prayer at this time. On Monday morning she underwent two operations—lasting in all about four hours. Both were serious, and the recovery is expected to be slow and painful. We saw her just after we landed, and she was radiant in faith and love toward Jesus. She is doing as well as can be expected.

Much lovingly and "dovingly,"

Frances

[The following letter was inserted without date and seems to be a postscript to the letter above, written after the Anniversary Meeting. It is included here for completion's sake.—ed.]

Again the Lord showed me that some dance often when alone with the Lord, and thus have more liberty and beauty in the assembly. Others do not, hence are not trained or exercised for the meetings. The Lord has called us to both interior and exterior life—and none of us can hide away from either life, and please Him. It is not easy to manifest His praise when we have no direct anointing or unction or even a feeling to do so. But, I praise God, as I really made an effort to obey, the Spirit wondrously helped me, and carried me far out of myself. I returned renewed and revitalized in a precious way.

There were many other interesting notes in the reports, but these will give you a composite picture of the evening, I am sure. For myself, I had a real struggle to be abandoned to the Spirit on this night. Various things seemed to hinder and distract me. I seemed to discern some were too aware of the cold and were fearful of becoming affected by it. I had to overcome the same fear, for if I sing in cold air, I usually lose my voice in a few minutes and have a sore throat. When Edeth started preparing the refreshments I was grieved that several went into the house and seemed too willing to end our worship and eat. So I forced myself to go back out of doors and continue in worship. I was also grieved that some had not sung or entered into the worship as freely as the Lord desired. I feel that we can easily fall into being an "on-looker," rather than a *participator*, if we do not obey the call of the Spirit in *each* instance. I also felt something about going out of the garden and into the highways and byways—which we shall do, as we know. I was on my way to the road when I suddenly looked into Chelo's "heavenly moon-struck" eyes. It seems that the Lord appeared to me in her eyes, and to her in my eyes, simultaneously. We started singing. It was a great effort at first. But I knew that if I were scheduled to sing in public before a crowd, I could and would do it, no matter how I felt. (I have done this out-of-doors a few times in the past.) So I did cooperate. And soon a spring seemed to open within, and then *I couldn't stop singing*. It went on for a considerable time. I didn't lose my voice, I didn't grow hoarse, nor did I get a sore throat. But if all this had happened, I would still be glad I sang and sang and sang. (Marian says I had a "soar" throat instead.)

I wish I could say that I felt the Lord had a full measure of song and love and praise from each one of us. He surely deserves it! This night will never come again, and it seemed very painful to me that some glided through it and missed the heart of the matter entirely. If this is taken for a lesson, then we can all profit by it. But to have to learn the same lesson over and over again is surely a matter of loss.

Several of you asked for the words of "On The Wings of Song" by Mendelssohn, which I was led to sing. Although it mentions the Ganges, we felt the other great rivers of the world as I sang—The Nile, the Seine, the Tiber, etc. (These words are all I can recall.)

> On Song's bright pinion ranges
> My airy flight with thee.
> Hence, to the banks of the Ganges,
> My dear one, O come with me!
> A garden we'll find that's perfuming
> The air with its fragrance rare.
> The lotus flowers brightly blooming
> Invite their sister there.
> The lotus flowers bright blooming
> Invite their sister(s) there.
>
> Beneath the tall palm trees sinking
> There we shall take our rest.
> From love's pure fountain drinking,
> We'll sing the songs of the blest.
> *From Love's pure fountain drinking,*
> *We'll sing the songs of the blest.*

In our *Anniversary Meeting* report I overlooked the dear song Elda was given for this meeting. She phoned it to us that Sunday morning, and it was shared at that time. We are including it now for the record.

> There's a time of the year,
> That we always held dear,
> Anniversary time with the Lord.
> It's a time we remember,
> Whether June or December,
> Anniversary time with the Lord.

cho.:

It's summer in our heart today.

It's summer in our heart.

Seasons may come, seasons may go,

But summer lingers in our heart.

Days may be drear,

But He always is near,

His kindness and grace He imparts.

And we always remember,

Whether June or December,

Anniversary time with the Lord.

Although Elda could not attend our Serenade, the Spirit gave her a little song to the Lord to sing in her "desert haven."

In my little "Haven in the Desert"

My Love has made His resting place,

Created beauty—lavish, lovely!

Where'er I look I see His face.

In the heart of gorgeous flowers,

In dancing leaves on vine and tree,

In the fountain playing softly

Love made a trysting place with me.

OUR FIRST MISSIONARY JOURNEY
INTRODUCTION

Our first missionary journey has been accomplished in the fulness of God's time and the amplitude of His space. For so long a time it existed only in the two-dimensional realm of promise and prophecy. Now it has passed into the third-dimensional realm of earthly performance and has been fully and perfectly recorded in heaven. So it seems fitting and pleasing to the Lord that I make an attempt to record it carefully on earth. This calls for a stirring up of my tired memory, which stored up myriads of sights, sounds and impressions all along the way; a reviewing of my fragmentary and sometimes illegible notes, hastily jotted down on planes, buses and in waiting rooms; and, most needful of all, the faithful assistance of the Blessed Holy Spirit—who has power to recreate past events, and bring to mind all that the Lord has said and done in this or any other time. All praise to Him!

It has pleased Him to give the trip this honorable title. I am aware that it in no way compares with Paul's first Missionary Journey, nor with that of any of the noble missionary-saints. But nevertheless it was truly ordained by the Lord, and the significance and fruition of it are known only to Him. It climaxed our two-year period of interceding for every nation under heaven, visiting many of these lands in the Spirit, sacrificing to send forth the Gospel to each one, and earnestly studying about them and their peoples, so that we might love and understand them and their needs more perfectly.

This call, while new in its implications and operation, was an extension of an earlier call which has been with me—and with all who have been in close unity with The Golden Candlestick—for many years, a call which is far-reaching and complex in its accomplishment. But it can be stated simply in a few words: we are to pray, believe, love, praise, give and obey to the *uttermost*, so that "The Gospel of uttermost salvation may be preached in uttermost demonstration of power to the uttermost parts of the earth." In the truest sense of the word, this is *"our"* calling. Just as truly, this trip became "our" trip, in which we all participated together with our Lord and His precious saints and angels.

Even earlier than this recent call to the nations, several of our sisters had strong impressions about going to The Holy Land. The thrilling news that the World Pentecostal Convention was scheduled to gather there in dear Jerusalem—where Jesus died for our salvation and rose again, giving us all eternal life; where the disciples tarried in one accord in the Upper Room until the Holy Spirit was copiously poured forth; and from whence the anointed ones went forth with the Gospel to the ends of the then-known earth—was thrilling and evocative. The very place! The very season! The very people upon whom the Latter Rain has fallen! The privilege to participate in such an event seemed very great. None of us felt worthy of it. But much prayer centered about it; and several began to make both spiritual and material preparations to send me, feeling certain that the Lord had appointed me as a delegate to represent us all at that auspicious occasion.

It was evident from the first that it would take faith, love and self-sacrifice to make this possible. I am sure that I shall never forget the devotion and generosity shown by those who went to extremes to provide for me, even though they truly longed to go themselves, and under some circumstances could have done so. A few received strong personal dealings along this line, and yet readily transposed them to me, realizing anew that we truly are "one"—and that they could go with and in me, by the power of the Spirit. I appreciate also those who accepted the announcement of my going by *faith*, not having been given previous dealings or revelations, and straightaway began to pray and sacrifice and stand with me in the Spirit. The Lord was gracious in giving each of you, as well as some others scattered abroad, a part in sending me and going with me, both in the material and spiritual realms.

It gives me much joy to know of the wide variety of dealings and experiences you have had both before and during this trip. Truly never did so many ride on one ticket! I wish I could write the record of this multiple visitation to the Holy Land and the other five nations. This, of course, is not feasible. Each of you must prepare her own report. And I am led to ask you to do so. No doubt only the highlights can be given, but do be sure to include the Scriptures given to you, since it is on the

Word that we stand and by it that we move forward. If you have faithfully kept your diaries, this will not be too great a task. Do this joyfully as a work of faith and labor of love, and make it a special offering to the Beloved. One month's time should be ample (dating from June 25.) Remember, we are called to be witnesses—and it is required of such that they make their testimony clear and true. We are also stewards of the mysteries—purposes and plans—of God, and, as such, must give a careful accounting. Please include that which you have already shared with me, to refresh my over-worked memory, and clarify my understanding.

I hope soon to "take off" into a "Dove's eye" view of the trip, accompanied by a very human "candid-camera" accounting of the Lord's testings and schooling. Were I writing only for those "scattered abroad," I should likely stress only the movings of the Spirit. But since each of you is a participator along with me of both the "kindness and the severity"—the blessing and the discipline—of the Lord, it is expedient that I share my own experiences with you, that you may profit by them, and perhaps hasten your own preparation and perfecting as a faithful witness. The words of Jesus must be repeated and manifested by each of us: "To this end was I born, and for this cause came I into the world, that I should bear witness unto the truth." Therefore, do bear patiently with me in this account until we are "airborne."

PROLOGUE

"God had prepared the people," and "the thing was done suddenly." Everything our Lord does—no matter how suddenly it happens—is accomplished only after careful preparation. With wonderful skill and patience the Lord painstakingly prepares us step-by-step for glorious purposes and consummations He has designed in His wisdom. Occasionally we are taken into a high point of vision, where we may survey the land over which we have passed. At such times we understand more fully the need of all the lessons, difficulties, sorrows and tests we have encountered. We are gratefully aware of the fruition of these in increased faith, understanding and consecration to the Lord. We overflow with gratitude to Him for our costly schooling. Then, as through tear-veiled eyes we catch glimpses of the land that lies ahead, *we realize how much further schooling and perfecting we need, to finish our course and enter with joy into our eternal inheritance.* My journey into the clouds—and far above them—served to bring about another such time of review and preview, and it reminded me in many ways of my mountaintop experiences in Sequoia during "Rapture"—in 1942.

There were several similarities! At that time, after a most supernatural and painful preparation, during which many prophecies and promises of God were manifested in various ways, I was finally ushered into "death" and "resurrection" life. The heavens were opened to me, and the Kingdom of Heaven was brought into such proximity that I actually lived and moved daily in the presence of the King. I was taken to the General Assembly in the New Jerusalem, in company with saints and angels; and when I returned I was instructed to be a faithful witness to the *Ecclesia.* I felt completely unable for so amazing an experience, and was dazed—even blinded—at times; was unable to hear or understand much that took place; was both humbled and uplifted; and was broken and refashioned in God's gentle hands. During it all—after the darkest and most trying parts were over—my heart's desire was that those who were near and dear to me in the Spirit might also enter into this heavenly realm and share these overwhelming experiences with the King in the Kingdom. And this desire, which was from the Spirit, was later fulfilled in multiplied ways. Even today *Thru Rapture* is still being used to open the Kingdom door and lift others into a new place of union with Christ.

In telling the story of Rapture, I skimmed hastily over the extreme trials through which I passed. The painful human and satanic forces which had to be encountered—the ever-present "negative" side of the story—was not put on paper. This unfortunately led some to assume that such supernatural experiences can be entered into lightly and inadvisedly and without self-death and complete subjection to the Spirit. In other words, even though they desired rapture experiences, they were not prepared for the dangers and sufferings that inevitably accompany such excursions "beyond the veil." So I shall not spare the telling of these as I recount this present journey into "death" and onward to the earthly—not the heavenly—Jerusalem, in company with angels and saints.

Everything God does must necessarily face strong opposing factors, and we must constantly encounter and overcome these in our own nature as well. Therefore it is highly important that we be instructed and thus forearmed, for none of us shall escape the "trial by fire," when God makes a real move in us. The more Christ is embodied in us, the more we shall encounter the open display of the powers of darkness. Only the Overcomers can hope to go forth with Christ—either physically or spiritually—in the glorious power of the Latter Days. *Unfortunately Overcomers are not formed suddenly in a high moment of Heavenly Visitation.* They are forged slowly in the daily furnace of affliction, and are put through extensive trials before they are ready to face the open onslaughts of the Evil One.

This Way of the Spirit is not for the faint-hearted or indolent. The fearful are always found with the unbelieving and are quickly overcome. How is it then that most of us—and certainly I—are inclined to be fearful, timid and backward, as well as lazy and slow to understand and obey? If we looked upon our own weak hearts and minds and bodies, we would simply collapse and give up, hoping to find some nice, comfortable church where we could salve our conscience and sleep our remaining few days or years. However, the facts are that we have already slept the sleep of death, and have risen again in newness of life! Whether we realize it or not, *none of us has a personal life to try to save and shelter,* for Christ now lives in us and MUST NEEDS find full expression for His LIFE OF THE AGES in this PRESENT TIME—in our weak earthly vessels. Thus we are in a strange dilemma! We are always bearing about this Amazing One who is full of dynamic love and faith, who knows no fear or doubt, who is seeking expression everywhere among all creatures. This means that fearful, silly, weak and utterly contemptible "I" must remain constantly—yes, hourly—planted in self-death and ignominy—not being heard, seen or noticed—in order that Christ may live and move and speak in me. As simple as this may sound, it is actually *the most complicated of accomplishments* for which we must strive daily. The method by which this is accomplished is two-fold: 1) by grace, through the continual operation of FAITH, apart from all feelings and appearances; 2) by constant self-discipline, in which our *will* must be persistently exercised—also in contrast to all feelings and appearances. If either faith or self-discipline is neglected and does not function, it is very easy to falter, fail and fall, or even to "make shipwreck" of our life.

Realizing all this from much past experience, you can readily see why I viewed the possibility of the trip with such sobriety and lack of enthusiasm. I have learned my way around in the Spirit—to a degree. I have been through the exercises and temptations and dangers of this means of transportation for many years. In the beginning, it is true, there was much I feared. The unknown is always hazardous: and our imagination, played upon by the clever enemy, can become overwhelmingly foreboding. Some have had mental breakdowns, or have come under dangerous deceptions after venturing in such realms. It has taken constant faith and self-death for any of us to continue on, year after year, in this hidden life of the Spirit, passing from realm to realm, without the comforting and stabilizing evidences of outward accomplishments. Yet, praise God, by the grace of Christ, and the constant assistance of the Holy Spirit, we have passed through crisis after crisis, and continued on in this high and holy way of living, worshiping, moving and acting in the Holy Spirit.

Throughout the years I have been transported many times to The Holy Land, as well as to other nations. It, of all places on earth, is real and dear to me. I was a little loath to visit it in the flesh—lest it would never be quite the same to me again. How much more perfect, pure and beautiful everything is in the Spirit, without the taint of the flesh and the senses! For example, The Mount is often more eloquent and sacred to those who come only seldom outwardly—but often in the Spirit—than of those who live here continually and have to see it in the common light of everyday. During the testing seasons, when the enemy is permitted to cast over it and use his deceptive shadows and images, he sometimes arouses in us an actual distaste for it and for these holy ways of the Lord. (If you have not suffered this sort of attack, do not let this statement trouble you. But be forearmed for such. The enemy is sometimes permitted to caricature, or misrepresent the Lord, His people and movings, and even His Words. And at such times actual distaste comes upon us. This can arise too from our own carnal nature, which periodically tries to assert itself and draw us away from the life of the Spirit.)

I cannot recall when the first intimations of an actual trip to the Holy Land began to creep into my mind. Certainly it was many, many years ago. One of the first songs my mother taught me was "The Holy City"—and as it was amusing to people to hear so small a child try to sing so magnificent a song, I was frequently asked to do so. It always stirred my heart in an unusual way—and Jerusalem became a most beautiful word to me. When I would hear people talking about how much they would like to travel to Europe or Asia, I would usually say, "Well, the place I would like to visit is the Holy Land." Yet I do not think that this trip came about through my own desires or thoughts, but rather by the direct will of God.

As early as 1941 the Spirit moved and spoke to us about the Holy Land. In fact, even before this time, I was drawn to the Palestine Exhibit in Los Angeles, and to Brother Futterer who directed it. On our tenth Wedding Anniversary, in 1938, Dwight and I were led to take ten couples there for "an evening in Palestine." A charming Arabic dinner was served to us, while we sat on the floor, or on low stools. I recall Brother Futterer placed me at the table under "Ruth's Wedding Canopy," and spoke words that seemed prophetic. I felt a sweet oneness with the Land, and the certainty that someday I would go to "the land of my fathers." After dinner we were conducted to the exhibit of art treasures and handiwork of Jewish, Arabic, Syrian and Egyptian origin. Then our "host" took us outdoors to a large illuminated map of Palestine, and gave us a lecture on modern developments there. He was very well-informed indeed, and had spent much time there, seeking unsuccessfully for the Ark of the Covenant, which he believed was hidden by Jeremiah on Mount Nebo. I later purchased his book, *Palestine Speaks*, yet I never studied it fully until I was preparing for this recent trip.

My prospective going to The Land was brought into focus by the Spirit every now and then. But always it seemed to be in the distant future, when Israel would be restored, and God's Endtime purposes would be hastening on toward Consummation. Even when the Spirit indicated that I was to go in May 1961, the date seemed premature to me. I wanted to postpone it. Yet I suddenly realized with a shock that Israel *has been in the land as a recognized nation since 1948.* And event after event in God's Endtime Program *has already been fulfilled!* At times it seems to us that God's time tarries long, and we spend our years in waiting. Again His time slips up on us unawares and takes us by surprise! So it was with me!

When I was certain that the hour had come to rise and go, I was most grateful for the long and careful preparation the Lord had made in me and in all of us. During the last year this preparation was intensified. Our following the footsteps of Jesus, reviewing His acts and words in each sacred place, was of great value. I found that my mind was hazy about biblical *places*. Yet God ordains both places and people, and sets them apart for honor. The map of The Holy Land became alive to me. And Dr. Field's book was a daily companion. (In fact I took it with me on the trip, since I could not possibly retain all its wealth of information and inspiration.) I read many other books in addition, making frequent trips to the library.

For many years I had felt the desire to learn Hebrew. Yet I could not seem to find a way to do so. I began to feel a strong urge to obey this impression. And the Lord showed me just how to go about it, by obtaining some excellent records and books. I was very weak and ill when I purchased these records. Humanly speaking I had strong misgivings—for I had heard that Hebrew is one of the most difficult languages. As I meditated about this, the Spirit came on me suddenly. (I was in the car at the time.) He began to move in me in a "resurrection" way, much as He did when He brought me up out of death into rapture. I rejoiced in the Spirit that Hebrew—for centuries an almost dead language, except among scholars, has now been resurrected, along with Israel! And I felt a new hunger to learn it. I wish I could tell you that the Holy Spirit wonderfully helped me and I picked it up in no time. But such is not the case! It became a daily discipline—in the midst of all of our other studies—to learn a few words and phrases. Marian felt led to study with me, and we shortened our lunch period, in order to find time to listen to the records. Again, at night, we had our individual study times after our own times of Evensong, memorizing, study, etc. By that hour we were both weary, and it was a test to be faithful to it. (However, there is always a compensating factor. After trying to learn Hebrew, memorizing the Scripture seemed *easy*.)

While I was still in the throes of Hebrew, I felt a strong urge to learn Arabic. This was an overwhelming thought—for Arabic is even harder to learn than Hebrew! And no available records could be found. In fact, even books were difficult to locate—suitable ones. But again the Lord led. Ten minutes sandwiched in twice a day provided a smattering of Arabic. (Yet even this small vocabulary proved invaluable in Jordan and Egypt, and was still put to use in Israel, among the Arabs, and in the Algerian part of Paris.) In the meantime, French also was being resurrected in me. I had three years of it in Junior High, about forty years ago, and I have used it a little since in singing. After I received the Baptism, the Holy Spirit spoke to me at times in French, even giving me little French songs—to my delight. (I truly love this language!) When I began to study Hebrew, I found that, for some strange reason, my mind would revert to French when the Hebrew word eluded me. Soon I had a modern French book and was crowding in an occasional session with it. (This proved to be of the greatest help on the trip, for I had to use it to be understood in our hotels in Egypt, Greece and France, and also among various ones I met at the Convention and other places. French, I found, is used even more widely than English in these lands.)

As you may well realize, trying to think in three languages, in addition to my own, proved quite a tax to my mind. (I even had a daily session in English, as I was teaching Jody.) At my age languages do not come as easily as in youth. Yet the Lord truly did assist me—and especially when the time came to use them. The point I am making is that it took a daily exercise of faith and discipline to obey the Lord in this part of the preparation. At times I was sorely tempted to give up—and to regard it all as a waste of time and energy. I knew the Lord could give me tongues supernaturally. But, as in so many other things, He didn't purpose to do so. Neither does He give us the memory of Scriptures perfectly without a discipline on our part. He has arranged everything to require both faith and works (in the sense of obedience) and, like St. Paul, we too must daily exercise our wills to overcome our own tendencies and to obey God's will instead.

I realize that the spiritual preparation was even more necessary and rewarding than the language study. Each of you played a vital part in this, both in the Studies of the Life of Christ, and in Zechariah, which, as you will recall, I was still working on up until two weeks before I departed. In reviewing your notes on this—though they had been prepared months previously—I found blessed inspirations and additional light. The latter parts of Zechariah meant much more to me after the amazing dealings we had at Passover, when Jesus, as King of the Jews, identified us more fully with Israel, and the Spirit moved repeatedly in us to act on behalf of the Jews.

Among other commissions, I knew that I was to walk in the land in the Spirit and Word of Zechariah, and bear witness again of the prophecies committed to him so long ago.

Our May-Time meeting was a fitting climax to all the studies, revelations and dealings. In the Word and the manifestations of the Spirit, and also in the anointings and the laying on of your hands, and the prayers offered, I felt a special anointing and equipment for the journey. For just as you showered me with lovely cards and expressions of love, in addition to your material gifts, so did you lavish upon me the precious impartations of the Spirit. And how we (I) do need "that which every joint supplieth."

Following this meeting I was busy taking care of our missionary letters and offerings, and answering Candlestick letters. There had been a number of responses to our general Candlestick letter—and most of these I answered briefly. Other business crowded my time. And of course there was the final shopping and packing. Over a period of time I had done all the preliminary things I could. But it seemed impossible to avoid a last-minute rush. In the midst of this, the Spirit led me to prepare a special Pentecost folder—which you later received—and to write a personal letter to each of you. This was a little surprise to me—as I felt I had already said my goodbyes. I guess these letters were a surprise to you too, and, I hope, a blessing.

On our Mountain,

Idyllwild, California,

July 7, 1961.

"Then returned they unto Jerusalem . . . and they went up into an upper room . . . and these all continued with one accord in prayer and supplication, with the women, and Mary the mother of Jesus, and with His brethren."

Beloved Ones in Christ,

It is a joy to be walking with Jesus on our own mountain again, after walking with Him on Mount Zion, the Mount of Olives, Mount Carmel, and other sacred hills in His native land. Our month of Pilgrimage was an amazing experience from beginning to end! It seemed all too swift and short to our dazed human minds, for it was difficult to keep pace with the swift means of the Dove as He bore us from place to place. The presence of the Lord was truly manifested in abounding love and grace, and in liberty in witnessing and praising the Lord openly all along the way. It was a delight to find so many of you traveling with us! I was aware of a precious communion with various ones of you, and I know your love and prayers ministered to me in precious ways. I seemed to be traveling in a cloud of witnesses, and at times the unseen ones were more real to me than the brothers and sisters who were with us in the flesh.

In all we traveled almost as far as around the world. Our vehicles certainly varied, ranging all the way from an ancient sailboat, which had to do a lot of tacking in a heavy wind to get us across the Nile, to an ultramodern Pan-American jet that carried us with "flying carpet" ease around the great Atlantic and on to Germany in six hours. During this flight we soared as high as 37,000 feet, and traveled around 700 miles per hour! This occurred on Ascension Day, and it was truly a celestial-like experience. The sun rose that morning at the startling hour of 2.30—New York Time! And all night long we were in a shimmering sea of luminous clouds. It was easy to imagine that the Dove was about to carry us onward and upward even to the Father's breast. A little song the Spirit had given me came to mind. It speaks of "Flying away and being at rest, singing and soaring, soaring and singing, unto the Father's breast." I spent that night on the wings of heavenly song.

But not all our flights were so thrilling and celestial. Most of them were made in old and crowded planes—and some were long and rugged. I, who do not like flying and had never been up before, had much exercise in faith and praise during our fourteen takeoffs and landings. It seemed that our lives were in jeopardy almost every hour, and we were in weariness and thirst, in heat and in much physical discomfort during most of the journey. I kept thinking of all the physical discomfort and stress Jesus endured in the days of His flesh, and of the sufferings of the Apostles and other missionaries who have gone forth in other nations. We Americans are so used to conveniences and ease that we have grown soft. "Arm yourselves to suffer in the flesh," and "Endure hardness, as a good soldier of Jesus Christ," are sayings that have little meaning to many modern believers. Another one is "In everything give thanks." We had reason to recall these words daily, and by God's grace we sought to practice them. Fortunately, the Lord had prepared me for these trials well in advance. I had read too of travel conditions abroad, and knew what to expect in the way of uncomfortable beds, scarcity of baths, strange and sometimes unhealthful foods, and the lack of safe drinking water. This last proved to be the most trying of all—for I was always thirsty—or so it seemed. Touring requires countless miles of walking—mostly up and downhill. There is also the heat, the dirt, and the difficulties with foreign languages and money. There are the beggars and the thieves—finding in "rich" Americans a legitimate prey. And there are the endless hours of waiting in crowded Airports—often without a place to sit—for Passports and customs clearance, or for the charter plane to arrive and get ready to take off. I truly thanked the Lord many times each day for the years of training He has given us in praising Him at all times and in all circumstances—in pain, weariness, loss, heartache and sorrow, and all manner of contradictions and vicissitudes. Without such training, I would have been in defeat all along the way, as were many of our traveling companions, which I regret to say. It was a shock to find that Pentecostal believers so easily became upset, and instead of making the journey in praise, some of them complained loudly every mile of the way.

There were other shocks too. The Lord sometimes sends us trials for which we are not prepared—and lets us meet them "head-on." The first big test came on the night before we were to depart. Out of the blue we received a shocking wire—

our flight had been canceled! At the last moment the U.S. Bureau of Aeronautics refused to clear our chartered planes for overseas flight. There was some technical requirement that could not be met. This meant that several hundred of us had no transportation to Jerusalem! Our reservations had been made months in advance. And we had to make connections with other planes and arrive in various cities in time to use our hotel reservations. Everything was thrown into an upheaval! Our bags were packed, our goodbyes had been said, the last letters were ready to mail. But now it appeared that we weren't going at all. The telegram said to remain at home and wait for a letter. A letter! I knew that it took three days for one to reach us from the New York office of the Tour people. And there was no assurance that any other arrangement could be made at so late a date.

I can well imagine that a lot of Pentecostal believers were ardent in prayer that night, as were we. I was "Ready to Go, Ready to Stay," as the old song says. But after all the Lord had said and done, and after the sacrifices made by our Company to send me, it seemed unlikely that the Lord wanted me to stay. He reminded me of a song He had given me recently—and it began to sing in my heart:

> Singing as I go,
> I shall journey on,
> *For this one thing I know—*
> *The Lord has bidden me to go.*
> For no matter how things seem,
> This life is but a dream—
> And joy cometh in the morning!
>
> After the darkness of earth's long night,
> Joy cometh with the morning light.
> So, singing, as I go, I will journey on,
> For this one thing I know,
> The Lord has bidden me to go.

The morning light did not bring us joy, for things were still uncertain, but there were precious assurances in the Word, so we joyed by faith. Our God is able to do the impossible, and there is nothing too hard for Him! After various phone calls, we learned that it was possible that one plane would take off that night, as per schedule. But it would be from another airline. If we wanted to take a chance we could go to Los Angeles and wait. We took the chance! And wait we did!—from 8.30 p.m. until 6.00 a.m. Another strange night! There was no place to relax or sleep. We expected our plane all night. Finally about 5.00 it appeared. (We learned later that getting a plane and crew on such short notice was practically unheard of—it was almost a miracle.) By this time I was quite willing, even eager, to get on anything that would start me toward Jerusalem. I had been awake then for 24 hours—and it had been a turbulent day and night. I laughingly thought of the old song which says that the Lord will not force us to go against our will, "He just makes us willing to go." I was!

Our plane was seemingly overcrowded and had to refuel in Oklahoma. By this time my husband and I were truly grieved at heart at the chorus of fault-finding and complaining going on around us. I could scarcely believe that I was with Pentecostal people. I heard no prayer or praise—no mention of the Lord. Instead almost everyone was blaming the Tour people, blaming the Airline, or even blaming the stewardess and pilot. For some reason the plane was slow in getting refueled. The wind was blowing a gale and almost everyone was worn out, hungry and upset. One of the women argued angrily with another, and I overheard the pilot disputing on the phone with some official. It was a trying hour and I had to really exercise myself to praise and give thanks. Once back on the plane the chorus of complaints got louder. Our lunch was as scanty as our breakfast. My heart was indeed heavy that we were unable to bear up under such tests with thanksgiving. It seemed like a marvel to me that we were going at all. But only a few seemed to share our feeling of thanksgiving. The chorus of complaining was finally silenced by a "fasten seatbelts" order. And soon we were right in the heart of a storm. Our pilot didn't have time to go around it—we were already late for our next plane! For a few hours all was oppressively quiet—except for the chugging motors—as we tossed around. Perhaps some slept, but I think most were praying. When the lightning began to flash a few even prayed out loud. I wondered if the Lord permitted that storm in order to quiet and solemnize us. It did!

By the time we landed at New York I felt ready to get on a train and head for Los Angeles. The thought of crossing the Atlantic in the storm was overwhelming! It was then that we were swept aboard the Pan-American Jet. How good of our Lord! Had we not been late in starting, we would never have been given this Jet flight. Instead of 12 or 14 hours across, we would make it in six. Finally at 11.30 that night we were given our dinner. And for a while the complaints died down.

I will not prolong the telling of other trials along the way, for you are eager to know the good things—the movings of the Spirit. And I am eager to tell you about them. We flew from Germany to Beirut, Lebanon, also on Jet. But again we were late—too late to get to Jerusalem, Jordan, Airport before dark, as it has no landing lights. So we were finally put on a plane for Amman, Jordan. I was thrilled to get to land in Lebanon, as I dearly love that land, and to have a chance to greet some of its people. I tried out my Arabic, but found French was more expedient. And it was also a joy to have a view of the lights of Damascus—with thoughts of St. Paul—as we circled around Amman, the capital of Jordan. I had difficulty praising during this flight, for I had slipped and fallen in the Airport at Beirut, and had lost my glasses. Also I had kept the whole planeload waiting for me. By this time I had been about 55 hours without sleep, for I could not sleep on the planes. And we had been about eight hours without a meal. Physically I was beginning to grow very shaky. I realized we were still a long way from our destination—Jerusalem. (And I had read about the road from Amman which takes one by the Jericho Road.) But when we landed, the Holy Spirit sweetly anointed me. And I had a blessed time witnessing to various ones at the Airport. I felt right at home, and the little Arabic I knew seemed to touch their hearts and soon I was surrounded by a crowd of Jordanians in native dress. I didn't care how long it might take us to get through Passports and Customs. How precious is the quickening power of the Holy Spirit! At last we were crowded—with all our luggage—on ancient buses, and began the jogging journey through Amman. But soon the buses were hurtling along in true Jordanian manner. It was breathtaking! We were flying now on the ground. (Fortunately I had read about how skillful these bus drivers are. This comforted me on the increasingly curvaceous, narrow road.)

The chorus of complaining increased now—for the way was very wearisome. And we had not been given a hot dinner—or indeed any at all—just a snack of bread and cheese. I was again exhausted and dazed, but suddenly realized that I *must* sing aloud God's praise going up to Jerusalem on the Jericho Road, as my spiritual ancestors had done centuries ago. So, by God's grace I did sing in the midst of the din around. And soon the Spirit came upon me in a mighty way. For over an hour He sang forth and moved in prophecy—in rejoicing and in weeping. He spoke about our coming days of Convention in Jerusalem, and how He had appointed this gathering for His glory and exceeding great purposes. He revealed that He would work in many hearts. But He also made it clear that there would be no open breaking through of the Holy Ghost, as many had expected. The Spirit wept because of the lack of praise and thanksgiving and real worship manifested among the Pentecostal believers. (Several times I had commented to various ones while enroute, "If we, the Pentecostal people, cannot manifest grace and praise in the midst of life's difficulties, then who can?") This same thought came forth now strongly in the Spirit. He compared us to Israel—murmuring every step of the wilderness journey. I tell you this with pain, and with no thought of condemnation of others. I felt identified with all Holy Ghost people, and acknowledged my own lacks too—as I journeyed with Jesus up the Jericho Road. I realized more clearly than ever how important to Him it is for us to give thanks and praise to Him continually—not just at certain times of worship, or in services and assemblies—in all things in our daily life. The Holy Spirit impressed me strongly that our Lord finds His joy and refreshment in the praise and ardent love of His people. I felt also His will that we not only give Him praise, but *become* His praise—a living manifestation of His glory.

All this was transpiring while the dear ones around me were talking loudly about a variety of things. Some were making fun of the bus and roads—in typical American Tourist fashion; others were complaining; some were jesting. Those who were being quiet were turning to look at me—as though I were disturbing them. I thought it strange that among Pentecostal believers, going up for the first time together to Jerusalem, song and praise should have seemed an offense. But it was so. Nevertheless I continued on in the Spirit. And later several told me what a blessing it was to them. No doubt a few were praising the Lord inwardly. But it surely seemed to me that, in such a group, our entrance into the Holy City should have been one of a spiritual nature. (However, by this time, we had been in our clothes for over sixty hours, and had slept little, if at all.) When at last the lights of Jerusalem were visible and some of its buildings came into sight, I could not restrain myself from weeping. I felt like a pilgrim indeed, and also very much like an Israelite coming up to the city to keep the sacred Feast of Pentecost.

From our hotel balcony we could see the Mount of Olives. And as tired as I was, I stood for a time on that balcony and worshiped under the stars. The city was quiet now, and only a few lights were visible—for it was well past midnight. But on the Mount of Olives a huge light—which appeared like a star—shown out into the night. I was enraptured! But I was also exhausted. By the time our suitcases arrived, it was past two o'clock. And one was missing entirely! I had some of our best clothes in it. I was tempted to feel dismayed—having already lost both my glasses and one suitcase. (However the suitcase was found the next day.) We were awake before six. Jerusalem stirs early. What biblical sights greeted our eyes from the balcony! We hastened to dress and go out into the streets. I felt the Lord was urging me to rise and run after Him in the dawn. Already the streets were thronging with workers, children, sheep and traffic! A short walk up the cobbled street brought us to Herod's Gate and the sight of the Dome of the Rock Mosque just inside the walls. But we hastened on until Gethsemane and the Kidron Valley were in full view. And just above was the Golden Gate, through which Jesus interred on Palm Sunday. (It has been walled up since 1540—for Moslems have a tradition that a Christian Conqueror will pass through it upon some Friday, and the city will no longer be theirs.) As we walked, my husband and I conversed with the people. We found them very friendly, even though they might know only one word, "Ha-low." (They have their own delightful accent.) I knew only a few phrases in Arabic, but they were eager to teach me. A few spoke English, and wanted to talk to us, and this was especially true of the boys. (Few women were on the streets that early.) We sang, praised, took pictures and truly felt that we were walking in old Jerusalem with our Lord. Their dress, language (so close to the Aramaic that Jesus spoke), their animals and hearts—all looked like something out of a Sunday School paper. Though we had enjoyed but little sleep, we both felt wonderfully alive and joyful, and yet we wept as we walked.

Often, during the days that followed, I seemed to feel Jesus' tender eyes look out through mine upon the people. And the Scripture which speaks of how Jesus looked upon the multitude with compassion came to mind. At times His love would fairly throb in my heart, and would flow forth upon those I met. Sometimes it was tears that overflowed, as the Spirit wept and interceded. Yes, Jesus is still weeping over Jerusalem—weeping through His Intercessors, by the power of the Spirit. And not only for Jerusalem and the Holy Land, but for all the nations of the earth!

Our time in Jordan was painfully short. Only two and a half days! With a Moslem guide, who was most kind and reverent, though hard to understand at times, we visited the famous holy places: the streets of the Old City; St. Ann's Church, where Mary is said to have been born—and the exciting Pool of Bethesda, only recently excavated after being lost for centuries. (John 5:1-16) My husband climbed all the way down the slippery descent—about forty feet—to put his hands in it and then put some on my head. (A portion from John 5:1-9 is displayed there in some forty languages.) Along with its excavation, they have also uncovered the remains of ancient churches and buildings on that site. It is amazing how many discoveries the archaeologists are making throughout the Holy Land.

Our Moslem guides took great pride in showing us the former Temple ground, which is so large that it contains one-sixth of the entire area of the walled city. Their own buildings stand there now. And no one can deny the beauty and majesty of the Dome of the Rock Mosque. The dome has recently been gilded, making it shine out beautifully. I stood there and tried to picture Solomon's more glorious Temple, with its great golden front—a dazzling site by day and a shimmering glory by moonlight. The Mosque *is* elegant! And while we reverently walked within it, the Spirit moved strongly in intercession for the Moslem peoples. He also spoke to me prophetically. And later I found surprising verification for what He spoke. When I saw all the old Wailing Wall—now barred to the Jews, who may not enter Jordan, I wanted to both cry and rejoice. For this wall actually stood there in the days of Solomon and in the time of Jesus.

We traveled down the Jericho Road again in sweltering heat—about 106 degrees. Stopping at Bethany we saw the traditional tomb of Lazarus, the place where the house of Simon stood, where Mary anointed Jesus' feet; and some partially restored ruins of an old house, which was on the site of the home of Lazarus and his sisters.

Our visit to the Jordan, Jericho and the Dead Sea was most interesting. But it was all very rushed. The altitude is the lowest on earth—reaching to 1292' below sea level. This, combined with the heat and the burning thirst most of us felt, made it an unforgettable experience. It was hard to recall all the vitally significant events that had taken place in this region. But we viewed everything with a certain awe, and climbed about examining the excavations of the ancient Jericho (one of the oldest, if not *the* oldest, inhabited cities on earth.) The Mount of Temptation is nearby. And it was here that Joshua won his first great victory. We saw the ancient spring, sweetened by Elisha and named for him. It is still flowing. Returning to Jerusalem, we all but collapsed—three rides over the Jericho Road in less than 24 hours proved quite an exercise!

The next day was Mother's Day, and it proved to be a memorable day I shall long cherish. We were up early and ready to start our Tour before the heat had overtaken the city. Our first stop was The Garden of Gethsemane and the Franciscan Church of the Nations now erected around "the rock of agony." It was crowded with tourists and "pilgrims" from other lands. Some of these were singing softly, praying or just walking about in meditation. Our time here was so short that Dwight and I returned at another time for a more personal visit. We were then taken to the top of the Mount of Olives, and viewed all the surrounding land, including Mount Zion and parts of the New Jerusalem in Israel. It was a most blessed excursion. And to hear our Moslem guide, in all earnestness, tell us how Christ ascended from that Mountain and would one day return to it was heartwarming.

Our next stop was Bethlehem. What a privilege to go to the very place where Jesus was born—on Mother's Day! After visiting Rachel's tomb and the sacred places in "The City of David," it was our joy to have a visit with our beloved Awni—our little Arab "David," whom we have adopted and are supporting in The Christian Approach Mission. We had a most blessed time with the children there. The choir sang for us, and I sang for them. And we all sang and rejoiced together. I felt led to adopt another child—this time a sweet little girl. It was wonderful hour.

Then back we went to Jerusalem for a late lunch and the afternoon tour. This included the Via Dolorosa, and ended at the Church of the Holy Sepulcher. Although I appreciated this ancient church, I believe, along with many others, that it is more likely that Jesus was crucified at The Place of the Skull, which is just outside the Damascus Gate. I think most of us felt a quickening as we passed through this Gate, and on out in search of the Garden of the Tomb. Our dear Brother S.J. Mattar is in charge of this, and what a message he gave us, as we sat there in full view of it, thankful for the shade of the trees and the fragrance of the flowers—after the stench of the crowded streets of the Old City. The Scriptures poured forth from him—and the Holy Spirit moved in precious ways. Song and praise rose up—and a real transformation took place. We forgot that we were hot, tired and hungry, as well as a little disappointed in finding so many of the holy places completely built around and so changed from Bible times. Here was a tomb that *might* have been the very tomb. There is much to substantiate this belief. In any case, it certainly reminded one of that blessed tomb in which Christ was laid. We were led to a view of the hill, that has the shape of a skull, at one end of the garden. We saw also the ladder that leads down to the underground church which has been discovered there, hewn out of the rock. Some of the men went down it—but it was too dark to see much. The centuries were swept away, and Resurrection morning was very real.

In the evening Brother Mattar and his dear wife took Dwight and me to the chapel they have established on The Mount of Olives. It is above the Garden of Gethsemane, and has a wonderful view of the city. The old part of the building is used for living quarters, and the Lord's people come and go. The new part is very simple, with many windows. And truly it affords a blessed place where Protestant believers can assemble to worship, or just to pray and meditate. The Golden Candlestick Company has a small part in the building of this chapel, we are thankful to say. Brother Mattar discussed with my husband future plans for enlarging it. Since Dwight is both a designer and a builder, he may be of some assistance to him.

The time for our Convocation in New Jerusalem was drawing very near by now, but we had to board a plane and fly away to Egypt. I would much have preferred staying in Jordan. But Egypt proved to be a most interesting and dramatic land. All sorts of things happened to us there, and a number got sick. I fell off a donkey-cart and painfully hurt my back. I praise the Lord for touching me and making it possible to go on. None of us wanted to get stranded there! Truly it is a land of spiritual darkness. And how great is the need for the manifestation of the Lord. It amazes me that any of the missionaries are able to be fruitful—so strong are the forces of the enemy. Yet even here I had much liberty in witnessing and in personal contacts. My Arabic was growing, and I used French, which is widely spoken. (Almost no one understood English—except for a few common words.) Three days here seemed like three weeks. Our passports all were mysteriously withheld from us—and most of us got very anxious indeed to get back to Jordan. (Of course one does not fly from the *U.A.R.* to *Israel!*) Back to Jerusalem, Jordan we finally went! And enroute we had marvelous views of the Suez Canal, the Red Sea, the Sinai Peninsula and much of the southern region of Jordan. This time we *flew* over the Jericho Road, and it was a mere nothing. I felt as though we were receiving the modern fulfillment of God's beautiful promise to bear Israel on Eagle's wings. How happy we were to know that at last we were enroute to Israel.

I cannot possibly describe how I felt when at last we passed through the Mandelbaum Gate, walked across "No Man's Land"—and were warmly greeted in impeccable English by a Jewish Customs official. Back in Jordan lay the Old Jerusalem, with all its precious and holy places, its dear Arabs and small groups of Christians. Jerusalem has a broken heart—it is a city cut in sunder! And I dearly love both halves! Yet in Israel I felt that I had come home. Here, as in America, a nation has been formed out of many nations. "One out of many." And their amazing vitality, hope and expectation is communicated to the visitor. Here is a people experiencing national "resurrection" life! And one feels very close to God's power as one journeys in that fair land. Of course the Spirit has dealt with us so much about the rebirth of Israel—both before and after it took place—that it is no wonder we feel that we are a part of it.

The Holy City was teeming with delegates to the World Pentecostal Convention. It was a real invasion of the Spirit! They wore no badges as yet, but of course they were easy to recognize. We were surprised to find that our room faced toward the east and gave us a beautiful view of Mount Zion. I was enthralled, almost overcome with joy, at last to that Holy Mountain. And it was the very first place we visited in Israel. The next day the Convention began. And with much excitement we gathered in the large *Binyanei Hauma*—the municipal auditorium—located near the Hebrew University. It rang with joyful greetings and rejoicing in a variety of languages. Our Israeli hosts were most gracious and warm. It was indeed surprising to find so genuine a welcome in the land that once rejected Jesus. Now in multiple form He was present in His Body. Included in this Body were a number of Jewish Christians, some from Israel and some from other lands. I particularly enjoyed fellowship with these. My study of modern Hebrew proved most helpful. And in the days that followed I had many opportunities to witness and to manifest love to the Jews on the streets, in the shops, and the hotel and wherever we travelled in Israel.

The Convention meetings were well attended, and the delegates numbered a few less than 3,000. Some 40 countries were represented. The U.S. had by far the largest number. The English-speaking people predominated, and the next in number were Swedish. So each message was given in English and interpreted in Swedish, or vice versa. Those not understanding these languages were seated in the rear or the balcony in groups. Each had an interpreter who gave the phrase in their own language at the same time interpretation was given on the platform. You can imagine that there was indeed a "babel" of tongues. But it was thrilling and exciting to hear Jesus talked about in so many languages at once.

I am sure the committee in charge put forth much prayer and effort in arranging the Convention. And the difficulties to be overcome were many. It seems inevitable that large conventions must devote much time to business matters, reports and other details. And also to speeches! Our times of worship and praise seemed all too short. And some who were eager to seek the baptism of the Spirit, or a refilling, were disappointed that no such opportunity was given. However I soon discovered that perhaps more was going on in the extensive lobbies than in the auditorium. Some, it is true, were just walking around, buying souvenirs, visiting and snacking. But here and there were small groups engaged in prayer, praise and seeking the Lord. And the Spirit was moving! I had precious contacts and fellowship with saints from various parts of the world. And new missionary contacts were made, as the Spirit led me to speak to various ones. Prophecy, praise and love flowed forth. The Spirit has had centuries of experience in going "without the gate" when there is no room—or time—in the "inn."

The Sunday Communion service was the high point to me. It was indeed owned of the Spirit. And Jesus was lifted up in a marvelous way. How glorious to partake of His Body and Blood in the very city that once rejected Him! We were with Dr. McKoy and his dear spiritual son, Emil. How the Lord has blessed them in their witnessing in Israel! The most effective work there seems to be done person-to-person, and in homes, rather than in churches. But several fine Pentecostal works are being carried on there, with growing success.

It was indeed painful to leave Jerusalem. But ahead of us lay a four-day tour which took us to Ashkelon, Caesarea, Netanya, Jaffa, Haifa, Nazareth, Cana, Tiberias, Tel Aviv and other memorable places. How sweet it was to follow the footsteps of Jesus, even with a Jewish guide! And when at last we knew it was time to take flight for Athens, we were comforted feeling that we could still follow the footsteps of St. Paul. This we did to Athens and then to Rome. It would take a large volume to tell you of our experiences and impressions. From Rome to Paris—we were there on the day our President arrived, and thrilled to see Old Glory so widely displayed and honored. Paris to London—our time was all too short. Then the long flight to New York and thence to California.

Since this letter has already grown long, I will conclude by thanking you for your recent letters. I appreciate your loving interest. And I want to convey to you the sincere love of each one of The Golden Candlestick. At present we are much stirred concerning the needs of the various nations. Our own nation is much in jeopardy. And the Lord has called us to stand in the breach and praise in faith of the fulfillment of His glorious purposes throughout the entire earth—and particularly in The Holy Land. St. Paul's inspired utterance at Mars' Hill, where we too so recently stood, still rings forth with trumpet sound:

"God hath made of one blood all nations of men for to dwell on all the face of the earth, and hath determined times before appointed, and the bounds of their habitation: *that they should seek the Lord* . . . and find Him." (Acts 17:26, 27)

Yours, rejoicing in the Apostolic Way,

Frances Metcalfe

THE COUNTDOWN

I t was while I was writing my farewell letters to each of you that I began to experience a strong negative attack about the trip. However, I frequently encounter hindering forces when I am working on letters or writings, so I was not really surprised at first. But the impression I was not going on the trip kept growing stronger. And it seemed very foolish to continue to write farewell letters, since they would not be needed. All the varied last-minute preparations seemed futile too—since I would be remaining at home after all. (I have had such attacks when I was about to leave on other trips, and sometimes I feel a sudden antipathy toward going away within myself. I could easily mistake these feelings for an actual premonition that I should not go—if I was not aware that I carefully sought the Lord each step of the way, ascertaining His will.) Also a sense of fear would sweep over me at times, and a darkness—as though the trip were a snare or a deception. The fear I could understand, for the Lord had made it plain to me that the trip would be hazardous and extremely arduous for the flesh. I had never flown, nor had I wanted to. And I am overly timid, I think, about any sort of travel except the familiar way by car—probably the most dangerous of all modes, but one I am accustomed to. I knew too about the danger of various diseases in foreign countries, having carefully read about conditions abroad. Unsanitary practices increase this danger, along with various infections carried by food and water. I had faced these things and felt prepared to have faith for a safe journey and return, or for whatever experiences the Lord might have in store for me. I will confess, though, that the thought of being ill in a foreign land at times oppressed me—but not for long.

Since it is our Christian duty to keep our affairs in order, one must give special attention to them before departing on a trip. This meant that I must see that my will was up-to-date, and that everything was left so that others could carry on in case I did not return, or my return was delayed. There is no reason to feel morbid about such arrangements, and I had no such test, for I have been brought face to face with death a number of times, and in a variety of ways, over a period of years. I find such little "deaths" helpful to the soul, and helpful in keeping me careful in my walk—"striving always to have a conscience void of offense toward God and man"—and detached in my attitude, so that I do not cling too tightly to earthly creatures, places or things, or even to my own life. However, the strong impression that I would not make the trip became oppressive, and increasingly hindered me in carrying out these necessary arrangements.

Little darts kept penetrating my mind. And I realized that it would take only a slight accident or illness to prevent either—or both—of us from going: a sudden virus attack can lay one low for days; or an infectious disease like measles, for instance, to which we happened to be exposed, could prevent our leaving. When one signs up for a tour, one cannot just postpone the time of departure. And a last minute cancellation on our part would have meant a complete loss of our investment. I tried to meet each of these attacks by faith and praise, but a strange fear settled upon me and grew more annoying by the hour. In my mind I had likened my flight to the launching of our first Spaceman, which was taking place at this time. (And truly it seemed as dangerous and drastic to me as his flight might have seemed to him.) I felt as though I too was in my final briefing and testing. And toward the end I seemed to be enclosed in some kind of spiritual or mental "capsule," waiting for the "countdown," to begin. When it did it turned out to be a lot longer than I had anticipated!

Just as our Astronaut had to make precise physical preparations for his flight into space, so was I instructed by the Holy Spirit to make mine. Each day I was to walk briskly uphill for at least 20 minutes. I also took various physical exercises daily, knowing that our journey would require some extreme exertions. I was led to cut my food quantity in half, so that I could shrink my stomach and accustom myself to eating only a minimum of food. I was also shown to get extra rest and to retire earlier—as the departure time drew near. This last part I failed to follow through as I should have—and I suffered as a result of my disobedience.

I practiced packing my bags in advance, since we had to hold the weight of our luggage to 44 pounds apiece. And I was reminded of how the Spirit has dealt with us about the preparation of the Bride. She wants her trousseau to be complete, yet she too must travel "light." It was not easy to carry just the right wardrobe and accessories for a month's traveling in a wide range of climatic conditions. And if one forgets a much-needed item, it cannot be quickly replaced at a supermarket or shopping center. It is therefore vital to be completely fitted out, and yet without excess weight or bulk. So well was I assisted by the Spirit that I found I actually had room and weight allowance to pack my autoharp in my large bag. And it was a delight to have it along on my pilgrimage. I was grateful that several in the Company had spoken to me about taking it along. I had thought it would be impossible, but it wasn't. I felt the Holy Spirit's interest and guidance in the details of my packing. And into my bags went tiny tokens from various ones of our Company. I added to these some significant things from our past. I felt I was to leave these as "souvenirs" in the lands we would visit.

Thus in every realm—spiritual, mental, physical and material—I did make careful preparation, by the grace of God. And how valuable this proved to be cannot be overestimated. I know that only grace took me through. But even for grace to operate fully, preparation must be made. This experience has shown me more than ever *the need to cooperate with the Holy Spirit and not neglect our times of preparation—however long and tedious they may seem.*

Eventually the last day to be spent at home arrived. And I still felt that I was not actually going. I found myself lacking in my last-minute duties, feeling very dead inside and as though everything were unreal—like a dream which was a bit unpleasant. Our dinner over, it seemed fitting to say goodbye to our sisters, and especially to grandma, whom I had not recently seen. Yet I kept feeling how silly it was, since I was not really going away at all. And, again, I felt I had already gone. "Perhaps, after all, I am only going in the Spirit!" This kept coming to me. Dwight and I went first to see Jennie and Gwen. And we were just about to leave when Marian arrived, the shocked and reluctant bearer of "very bad news." I could scarcely grasp what she was saying, even though she spoke in plain English. The flight had been canceled, she said. The U.S. Aeronautics Bureau had refused to clear the chartered airline which was to carry us to Amman, Jordan. A telegram had been sent to inform us, and to instruct us to stand by for a letter. A letter! Our Travel Bureau was in New York, so a letter from there would mean a wait of three days! This was just incomprehensible! I had thought of almost everything that *could* happen—including nuclear war—but had never had a hint that the Travel Bureau might cancel out. We hastened home, feeling as bewildered as ever we had in our lives. The telegram held no assurance that any other means of travel would be provided. So we called Norma at once, to see if she could find out anything additional from Angelus Temple. She was as shocked as all of us were—just unable to grasp it. But she quickly rallied and agreed to help. As the word spread around, each one shared our concern, and I am sure fervent prayer was made.

When I went to the Sanctuary alone to seek the Lord, I had only one thought in mind: "What will this mean to the Company, and especially to those who have seen so much about this trip, and have made such sacrifices to send me?" It seemed an almost unbearable test—if I could not go after all. It was not a matter of personal disappointment, but of the seeming negation of all that the Lord had revealed and spoken. I realized that it would be very difficult to believe many other things He has said, if this proved to be untrue.

As shocked as I was, my mind became very active in recalling past experiences when I, or we, came face-to-face with sudden reverses, unexpected blows and griefs. Undoubtedly the Spirit was reminding me of these crises in order to hearten me—for I realized clearly that, in spite of all we have suffered in such contradictions and reverses, *the Lord has always given us grace and has moved on relentlessly toward His appointed purposes.* All praise to Him! There is a rich residue of faith deposited in my heart as the result of many past trials. And this seemed to revive me. Soon, I felt a blessed sense of personal peace and trust in this latest crisis. I realized also that this peace and trust was sufficient for all you dear ones too. I could rejoice that if I did not go, you would not be disheartened, but would still rise up to praise and trust the Lord, and would continue to run the race with faith and love. The devotion, sacrifice and effort you had spent in trying to send me would not be unrewarded. No! It would find fulfillment and reward in some other way. All our blessed months of study and preparation would also have a blessed fruition in the Spirit, regardless of whether I went in the flesh or not. In that hour I was positive—as I am now—that *to go in the Spirit is far more important, and sometimes more real, than to go in the body.* For days I had sung an old song, the chorus of which says:

> Ready to go, ready to stay,
> Ready to do Thy will,
> Ready for service, lowly or great,
> Ready my place to fill.

And I still felt that way. It was just the suspense and uncertainty that was trying. As I was thus in prayer, I suddenly thought of the hundreds of other passengers who were likewise disappointed this night. Surely they too had made much about the trip, had felt certain they were to go, and had made sacrifices toward that end. I realized that many fervent Pentecostal prayers were rising to God with mine. Yet I could not "storm" heaven to open the way to go—but only thank and praise the Lord. After a few more moments, it suddenly came to me that I hadn't thought about how *Jesus* felt at all! I had thought first of you dear ones, secondly of other precious believers. But of the Lord I had not thought, except in a general way. I was pierced in my heart and said, "Dear Lord, forgive me, and show me how YOU feel about this. Will You be just as happy if I do not go, or will You be grieved? Is it Your desire to walk with me there in the Land, in the body, or only in the Spirit?" He did not answer me, but the Spirit made it clear that I should not have been going in the first place had I not been certain of this. Had not He spoken over and over to me directly through various ones of you—and also in other witnesses—that He greatly desired me to go? Then, if so, what had altered? Had God suddenly changed His mind? And did the failure of an Airline cancel out God's will? Of course not! Then, there was only one thing to do—and that was to go. If one Airline could not fly, another one could. It was up to me to have faith that I was going—and to do everything possible to bring it to pass. I was overwhelmed with the thought of what all this might entail, but I rose from my knees knowing that I must go upstairs and shampoo my hair, and get to bed and to sleep, so that I could be ready for my next step. As I came up the stairs my little chorus sang inside sweetly:

> Singing as I go, I will journey on,
> *For this one thing I know,*
> *The Lord has bidden me to go.*
> For no matter how things seem,
> This life is but a dream,
> And joy cometh in the morning.
> After the darkness of earth's long night,
> Joy cometh with the morning light.

So, singing as I go, I will journey on,
For this one thing I know,
The Lord has bidden me to go.

During the night, though still tossed about at times, I felt FAITH "take hold" for the trip. Yet the thoughts of trying to get transportation and lodging at the Conference were very tormenting. I knew it would cost almost twice as much to go as individuals, rather than on a Tour, and at such a late time it would be almost impossible. But one of the sisters had said to me, "Remember the prophecy at our last meeting about God being the God of the impossible?" And I seemed to rest on this word—that He is the God of the impossible. I did sleep fairly well, but woke early, feeling very flat and tired—and hoping that this was only a bad dream which would now be over. But the "dream" was still there. So I had to rise and exercise faith and praise. My time in the Sanctuary was odd—and I could not feel at all that I was leaving it, or the mountains. Dwight seemed very heavy, and suggested that perhaps he should spend the day putting in the rest of a rock wall! I was amazed, and said, "Why, you can't do that, we have to get ready and go to Los Angeles. If we don't go on the Tour, we have to arrange some other way to go, and it may take a lot of time and running around." He said, "Do you really think we are going?" And I said, "Of course we are going." Yet the thought of the cost of plane fares staggered me. Our Tour company would have to refund our money, but that might take weeks. In the meantime, I would have to borrow. I recalled that two different ones in the Company had offered to loan us money, if we needed it. I wondered now if we would. In any case, we were to have some word from Los Angeles by ten o'clock, through Norma. And I recall that she greeted me on the phone with these words, "Every possibility!" It seems that the Tour people and Angelus Temple workers had been up all night trying to make other arrangements. They would have a confirmation about the plane within two hours. In the meantime we were to stand-by, and be ready. Dwight was now bright and joyful and announced that he was going to start the final packing. Marian and the others were most helpful, they exercised faith and praise and were not daunted by this test. Their confidence was surely a blessing.

But by noon no confirmation came. Nevertheless I paid a brief visit to the Sanctuary, to the outdoor sanctuary, and the little shrine. And I bade some half-hearted goodbyes. It was still very unreal that I was going on a trip. But on this strange day *our spiritual past* blended with the *present* in an amazing way—and all seemed to converge in me and point to the *future*. I felt particularly the trip I made to San Francisco to the first Assembly of the Nations—which later became the UN. How momentous was that day! I painfully recalled how, after many other tests had been waded through, Franklyn had suddenly broken out with a red rash and had a fever. It appeared to be Scarlet Fever—which was in the neighborhood at the time. Yet I was obliged to leave him all alone in the house—the other children being away. He was in his teens, but it was indeed a death to leave a sick child—even for only four days. (However it was only a light recurrence of measles, and he did fine.) He had begged me to leave—feeling God was sending me. And again things were timed exactly—for on this day, when I must leave for the Holy Land, Jody broke out with a red rash and was running a fever! She too was cheerful, as he had been, and urged me to go and try to get on a plane. But it was painful to leave her home in bed, after she had looked forward to seeing us off in L.A. (Ila was a dear and came in and risked measles to take care of her until Marian returned—but neither took the measles, praise God!) Of course by now my throat was feeling sore and I was feverish too. I began to calculate—how long do measles take to incubate? (Yes, I have had them. But they do recur.) Ah me! Again—Faith!

At length we actually got in the car and drove away. Our last word from Norma was that the plane to take us was not yet "confirmed," but they were working on it. We were to come and take a chance. It was a hot day, and we made several stops enroute. By the time we reached the Airport we were already tired and ready for a good rest. The "Countdown" had begun! They assured us that there *would* be a plane, but were not sure *when*. I was so thrilled and happy that a plane was promised—that I was willing to go on just about anything, in order to get to Jerusalem. However, my feeling of joy was dashed suddenly when I read the headlines of the evening paper: 75 had perished in a plane crash in the Sahara Desert! Up until that moment I had not dwelt on my fear of flying. I had always said that I would never fly unless the Lord told me to. And if He did, He would give me grace. As I tried to analyze the sick feeling that swept over me, upon seeing the headline, I felt it was not a fear of death—for meeting the Lord would be a joy. Usually a plane crash brings sudden death. But occasionally there is a lingering ordeal—and of these I had read. I think I feared crashing in a desert, or mountain or wilderness area, and being among the survivors, rather than the dead. Especially did I fear a landing on water. I had read of how some of them had floated around on rubber rafts for endless hours. The terror of lonely heights, depths and breadths is very real. And I recalled how *survival* have been given to us as a key word for this year! Yet for all such hazards I knew I must be *willing*. Ready to go, ready to stay, ready to die, ready to live, and ready for the *unexpected*.

On this special night being ready and willing to *wait* was what counted. The waiting room of Los Angeles International Airport is not equipped for a comfortable night's lodging. They are now in the process of constructing a very modern Airport, but the one we had to use is stuffy, noisy and crowded. But exciting! Yes, I guess all airports are exciting—for many real-life dramas are enacted there daily. I had ample opportunity in the days lying ahead to become acquainted with airports. But I little dreamed on this first night of nights how much time I would be spending in such a locale. At first we had company in our waiting. Marian, of course, was as thrilled as though she were going too—for she felt that she was. And soon others arrived, Isabelle, Gertrude, Kay, Rosemary and Naomi. I was indeed surprised that they came so far for this occasion. I had urged everyone not to make the trip, due to the uncertainty of our departure time. Soon Norma and Lamby, Chelo and Majella arrived, so we formed a little circle in the middle of the lobby, and doubtless blocked traffic for a time until we realized it. Orma also appeared and surprised me with a little shower of gifts—including some delicious health candy, which surely came in handy on the trip when meals failed to materialize.

By this time the Airport was crowded, and I was scanning the ground for other familiar faces—and wondering how many of the people would become our traveling companions. I was aware of the Lord's presence, and yet there was also a little nightmarish quality to this night, and there were several little trials—like the kind we encounter in dreams. For instance: the lovely corsage some had felt to provide for my "going away," which seemed to them most bridal, became a casualty when it was unwittingly frozen, in an attempt to keep it fresh. Such an unlikely occurrence—but it happened! Another corsage, however, was hastily provided, and much appreciated by me. Then a dear sister was introduced to me, who proceeded to talk at length and to bring other sisters to meet me. I was so tired and excited by now that I could not remember any of their names. And it was annoying to be more or less "taken over" by a stranger at such a time. I struggled, trying to be polite and loving, and feeling strangely like I was in a bad dream. (This dear woman persisted in this way until she left.) We also had a tormenting few minutes when we suddenly realized that when we had given our bags to the agent for weighing in, we had allowed both our overnight bags, which should have been kept with us, to be checked. This would mean that we would have no access to them until we got to Jerusalem. And there were things in them we badly needed. So, Dwight went out to the freight loading platform and engaged them in a search, which was annoying to the men and to him. They would allow only mine to be taken out, since his was too large. And in his bag were my comfortable shoes and other necessities! I was truly dismayed for a few moments by this inconvenience, and humiliated to make others so much extra work.

Yet nice things kept happening too! Out of our "past" Lydia suddenly appeared. She had found out about my trip, and had been much moved upon by the Spirit. It was difficult to talk alone with her or anyone, since all of us had to remain standing—there being not enough seats. However, I did take a little walk with Lydia and listened while she mentioned various things out of the past—most of which I too had been feeling—and shared with me the Word and impressions the Spirit had given her. Later on, after all our dear ones had left, at our request—for we realized that the plane might not be ready for hours—another witness from the past arrived—Cliff. By this time Marian, Dwight and I had gone over to the café for a hot drink. It was about midnight, and I had finally remembered that I had some comfortable shoes in the car. With these on, and a chance to sit down, I was just relaxing and feeling the Lord's sweetness, when this witness arrived. I laughed, recalling how often he had appeared at one of our homes at midnight—or thereabouts—in the "old days" when we were living with various ones of the Company—during the housing shortage. It was typical! And like the Lord too, who often comes at this hour. On this night he

49

had returned home late from an engagement, had phoned the Port and found we were still there, so had felt led to tell us goodbye. He too spoke of my trip to San Francisco—at which time he had been an anointed witness in our midst—and other past things. And he gave me a rosary which his sister had brought to him from Rome, asking me to carry it up the Via Dolorosa, and to remember him as I did so. As he handed this to me, a sweet assurance came to me that I would indeed reach Jerusalem and make that pilgrimage. It was comforting!

Soon he too had left, and we settled down in the waiting room to try to rest, thankful for chairs even though they were hard. The radio was playing, and all around us people were talking. It seemed odd to me not to hear the Lord mentioned, even though I knew many of these people were going on the Tour. Some were reading newspapers and magazines. Others were trying to sleep. Those still talking were expressing doubts and disturbances about the long wait. In the midst of this discord I had a sweet sense of the Lord's presence and His peace, as I attempted early on to joy in our dilemma—and praise Him. And I actually did not feel too tired or shaky. From time to time Dwight brought us little bulletins about the plane—and its expected arrival kept growing later. Some felt it would not come at all. Finally Dwight told us that the plane had arrived and was being serviced. He had found our exit and led us to it. Already others were gathering. And soon there was quite a crush there, as people pressed in from behind. The clamor was by now a real chorus, and I was truly amazed and almost alarmed. It was difficult to believe that we were in the midst of God's precious Pentecostal people. True, there was no swearing, no smoking, no drinking in evidence, but upset dispositions were surely manifested on every hand. Marian was as disturbed as I, and said, "I surely pity you if you have to travel with some of these people." Our *long* wait at this gate was the most trying time of the night. Tension kept mounting and many seemed to seethe in resentment. It was most difficult to keep our mind on Jesus at this point, and not to become upset—as we sometimes do—at those who were upset. Carnality is very contagious!

At long last the gate opened and I took a fast look at our plane. It was smaller than I had anticipated, and when I saw the crowd teeming to get on it, I wondered how it could possibly hold so many, and their baggage. I also noticed that it was The

Trans-Caribbean Air Line—of all things! Surely we were not going anywhere near the Caribbean! But things with the Lord often seem contradictory. I can assure you there was no time to ask questions—I was shoved forward by the surge of the crowd and found myself at the runway of the plane, looking up into the morning sky, now bright in the dawn. Just above the plane was the last quarter of the moon, looking like a new moon, and beside it, in radiant glory, the star of the morning! What a sight! "After the sorrow of earth's long night, Joy cometh with the morning light." Indeed! With joy I clambered up into the plane. We squeezed into a rear seat and waved and smiled bravely—if a little wanly—while the plane rapidly filled. The door was closed, we fastened our seatbelts, and the plane started its motors, one by one. We slowly taxied down the field in full view of the rising sun—waving to the last. Ten, nine, eight, . . . three, two, one,—the long countdown was finally complete, and the moment of takeoff had arrived!

THE TAKEOFF

T he first time one flies is a unique event likely to be long cherished in memory. It should be a real occasion—and it was! At just 6 A.M. our plane was in position and the last racing of the motors set us all to vibrating with expectancy. Dwight and I held hands, lifted our hearts and voices to the Lord, and felt the rush of speed and soon the gentle lifting of the wings. "Singing and soaring! soaring and singing!" This is the way we rose up on that morning of May 11th. I, who do not like to look down, forced myself to watch the receding land and then the blue sea beneath us. I was thankful indeed that I had read some about flying east, or I think I should have panicked to find myself winging my way out over the blue Pacific. As it was, I knew we had followed the usual takeoff pattern at Los Angeles and would soon be circling around and heading east. I greeted the sea and saw Catalina at a distance. (Last summer when Jody had begged me to fly there it had seemed almost overwhelming. Yet now here we were near it in just a few minutes!) I hailed all the islands of the sea, and the dear Orient, which I could not visit at this time, and then felt the swaying of the shifting plane as it began to bank and turn. With a little shock I realized that we were now about to cross the entire continent of the U.S. "From the mountains to the prairies to the oceans, white with foam." (Yes, God bless America! I would pray for each state over which we passed.) After crossing our own country, we must cross the vast Atlantic ocean, then the continent of Europe, the Mediterranean Sea, and part of the Middle East before we reach Jerusalem! The word uttermost came to mind. For it is said that from Jerusalem, following a direct line, Los Angeles is the "uttermost city" to the west. And, likewise, in an imaginary line to the east, one reaches Los Angeles as the uttermost city in that direction, halfway around the world. This may or may not be exactly true, but it *is* poetic. And are we not the Uttermost Parts Missionary Fellowship?

I had hoped to see some of our California cities enroute, but we were now flying at about 12,000' over some very rugged and unfamiliar terrain. A few of the peaks we

55

seemed almost to skim over—so high and near were they. I kept telling myself that I was enjoying it, for some had predicted that I would just *love* flying, as they do. I was hoping so! An old song came to me, and I sang it softly. The chorus says:

> Over the mountains, plains and sea,
> Here I am, O Lord, send me.
> I'll go to the ends of the earth with Thee,—
> Here I am, send me!

This song took me back to 1936 and '37—and the special moves of the Spirit then. I had been sure that I was actually going forth with Jesus soon—and had been for days and weeks in the wonder of it. I had sung it almost daily.

> Hast Thou, O Lord, a work to do?
> Here I am, send me!
> The fields are white, the laborers few,
> Here I am, send me!

Now, after 25 years, I was actually going forth over the mountains, plains and seas, even to the ends of the earth. Did not Abraham's trial of faith continue for some 25 years after he received the promise? Fulfillment! But of course I knew that this present trip was only the beginning of His more outward movings.

My meditations were interrupted at this point by the serving of breakfast. We were first to be served, being in the rear. We were now supposed to unfasten our seatbelts and "make yourself comfortable"—a euphemism for "make the best of a tight situation." The colored sister who was in front of me had immediately let down her seat, preparing to sleep—this placed the back of it near my lap. I was wedged between her and Dwight on my right. The seats were close together, and extra seats had been added on the aisles. Nothing heavy may be placed overhead on the racks—so only our hats and coats were there. Around us, on the floor, were our two bulging flight bags, Dwight's two camera cases, my overnight bag and my purse. I had taken my corsage off and pinned it on the curtain of our little "peephole"—so it wouldn't get crushed. I think I felt a little like an Astronaut fastened in a "capsule," for the seat of the colored lady wouldn't go up, even though the hostess and others tried to raise it. (I found I had a real penchant for getting behind such seats on other planes too—and if there was a sticky seat onboard, I was sure to get it, or get behind it!) But someway the tiny tray was squeezed onto my lap and I gazed into what the hostess brightly called "breakfast." It consisted of one small glass of weak orange juice, one small cup of weak coffee, and one small piece of pound cake. Yet it seemed delectable to me. As I began to eat, I couldn't help but laugh inwardly at myself. My idea of the spacious interior of planes and luxurious meals had been formed by T.V. pictures. And evidently I had seen scenes of "first-class" lounges. We were all simply jam-packed!

By this time we realized that we had made a mistake in the selection of seats— we were right by the two restrooms. As narrow as the aisle was, it was usually lined up two abreast, both men and women, waiting to get into the restrooms—which proved to be not "his" and "hers" but—to my dismay—"theirs." (As was the case abroad in some places.) As our brothers and sisters waited in this "lineup" they conversed—mostly about our long delay and the old and crowded plane, which was "noisy, crummy, antedated, and unsafe." As soon as they had eaten their breakfast, it became the next lively topic of complaint. Our hostess was from Puerto Rico, a charming and gracious little lady. She explained that since the plane had been sent on "emergency orders," they had not been able to arrange in advance for our breakfast. The L.A. Airport had not wanted to cooperate with this unchartered flight, and they had found it difficult to obtain any food at all—at so early an hour in the morning. She apologized over and over, to passenger after passenger. She was sorry we had been up all night, she was pained that we were hungry. I marveled at how patient she was, and was ashamed of many of our companions. I was getting to know them better, for sooner or later everyone on the plane was at the restroom doors. I tried my best to tune them out and keep my mind on the Lord and the trip. I was still feeling very thankful. I had been up for over 24 hours by now and was feeling fine. My breakfast seemed ample, for I was not hungry. But I was getting thirsty. I had no idea how to get out of the seats to get a drink, for our aisle was already wedged! Finally, by a sort of rude—or heroic—lunge, Dwight got out and up the aisle to the drinking fountain. The water proved to be lukewarm and L.A.-ish in taste. But any water was most welcome.

By this time we were flying over the Colorado river near Needles. And I thrilled with joy, seeing my native river again, and remembering how we had followed it to its silver-slim source high in the Rocky Mountain National Park. Now we were in Arizona, and I began blessing Elda and Ada and others there, and praying for that dear state. I was delighted that I could see Oak Creek Canyon so clearly from the sky—and the beautiful red rocks and formations that lie at its western end. Soon I knew that we were near Sedona, and its lovely pink rocks and buildings. Then came the Painted Desert, the Petrified Forest—and somewhere along there Holbrook and Elda. More buttes appeared and we were at Gallup. In a few more minutes Albuquerque was just beneath us—a thrilling sight! We had passed the southern end of the Great Divide—and there were the mighty Rockies and Colorado to the north! The flight was delightfully smooth and I was beginning to really enjoy it, and to gain a little confidence. Yes, I might make a flyer after all! But of course I knew it was grace that was upholding me. Our pilot was most cooperative in pointing out places of interest. But it seemed hard to find much of interest in the part of Texas over which we were passing.

By the time we reached Oklahoma City it was noon or after. (Of course we were losing time as we went.) I was delighted that we were to land there, for by this time I had sat so long in a cramped position I was longing for some chance to move and stretch. I knew it was hard on Dwight too—and all with long legs. I also wanted a change of atmosphere. The air on the plane was quite fresh and adequate—I am so grateful for the individual outlets that make it possible for each one to control his own flow of air. But the spiritual climate was anything but refreshing. Around us were seated mostly colored people. They, at least, did little complaining, or talking about anything. The white woman across from us was rather sick and disgusted with her tight seat—much worse than ours. When the "fasten seatbelts" sign flashed on, the aisles were cleared and quiet reigned. We were now about to land—and the airport looked very tiny and countrified from the sky. I hoped we would someway find the runway, for I recalled that the takeoff and the landing are the two most dangerous times—and once again I was exercised. The plane tossed a little in the heavy wind, but it smoothly glided down, touched and taxied in. My first ride was complete and I had survived! Wonderful! Nor had I been sick, nor unduly frightened. I think I felt a little like a puppy who is just ecstatic to find himself alive after a mean old bath. I wanted to run and jump and shake myself and "bark." But for a time I just sat still. When the plane is landing, off goes the air conditioner and it becomes stifling in just a few minutes. It seemed such a long time before the door opened and the passengers were out. We had to be among the very last, of course. Outside the wind was blowing very hard and my hair was tossed every which way. I was thankful for a tight fitting suit. (We learned that Oklahoma had been going through constant high winds and hurricanes—nearly 400 this season.) We were told to remain in a limited area near the plane, since a quick servicing and takeoff were anticipated. Several wanted to walk a distance of about two blocks to the café for lunch—and some did this, in spite of being told not to. We, of course, obeyed the rules and walked about. I felt to walk alone and have a "noonday" songtime. But walking in that wind proved tiresome after a few minutes, so I went to the little waiting room. There to my delight I found a fountain of ice water—and I took frequent drinks, walking about between them—trying to praise.

In the waiting room itself several arguments were going on. The pilot, on the phone with the Tour people, was mad about something and was heated in conversation. The hostess was explaining to someone that they were doing their very best on such short notice. One woman was really displaying carnality and self-pity, and was very ugly in speaking to our Tour leader, Leta Mae Stewart, from Angelus Temple. She, in turn, was answering her most lovingly and graciously. I truly loved this sister who was both beautiful and spiritual in her attitudes and actions. I knew she was greatly loved, both as a church secretary and a Bible Teacher, for I had heard her highly spoken of. I felt a great love for her and also for her traveling companion, Miss Hall, who happened to know my sister-in-law. I had conversed with them both on the night before, and had hoped to have some sort of fellowship with them. I went to Leta Mae at this time and told her how very thankful we were for all she and others had done to make the trip possible, that we were grateful for any plane at all—after this unfortunate cancellation, which they could not help. She seemed pleased, and spoke of how we should be thanking the Lord and not complaining.

But she met each new complainer with quiet grace and sought to comfort them. One woman said that the Lord had given her the verse, "In everything give thanks." This brought an eruption from our chief complainer—a dominating red-haired woman who was traveling in style, with a copy of *Glamor* as her companion—she said God would never treat her the way this Tour "bunch" was treating her. That God wanted her to have the best, because she served Him so faithfully. This made some of us gasp. And when another sister sought to tell her of St. Paul's faithfulness and sufferings, and of what he said about rejoicing, the complainer said that the Bible meant no such thing—and began to holler at us all. It was rather unnerving, for we were crowded in a small restroom by now in another lineup.

Suddenly, I began to feel shaken and out of place, and I did not want to get back on that plane with so much strife going on. I began to realize that no one had led in prayer before we flew. No one had sung or praised the Lord. And I had done so only quietly. Yet even my quiet praise seemed to irritate some and they eyed me coldly. Others asked me the stock question, "Are you with the Assemblies or Foursquare?" (As though they were the only two that counted. But actually many organizations were included in this party.) When I told them I belonged to "neither, nor"—well, I was aware of an instant withdrawing. They had never heard of The Golden Candlestick Church—and obviously showed their disapproval of me. Dwight didn't help matters by announcing—when questioned—that he wasn't actually Pentecostal at all! We were marked from the first!

I sought out a few sisters, during this long noon day wait—for such it proved to be—and tried to establish some sort of communion, looking to the Spirit to lead. But for the most part they were too tired, hungry or distracted to seem to want to talk about the Lord. There were a few exceptions—a few radiant, calm and seemingly grace-filled ones. Yet, on the whole, I felt out of place and uneasy. By the time we were back on the plane, we were truly hungry—it being long past the lunch hour. Everyone expected a hot lunch soon after takeoff. We left Oklahoma and soared over Arkansas. How beautiful and green it looked! One could see traces of the recent floods, however, and this was distressing. I compared our dry desolation with their wet desolation—and wished nature were a little more orderly with the weather. And I thought, of course, of John, Ila, Eddy and Grandma. I recalled that Oklahoma had been Edeth's and Argen's state. And later Tennessee brought sweet thoughts of Clement and his poetry that began to flow there.

However, lunch was being served and we were busy with it. But not for long! Again a tiny tray arrived with a small cup of coffee, a cheese sandwich, an apple, and another piece of pound cake. This really stirred up a tide of reaction. Surely we were going to have more than this after so long a wait! It was now about three o'clock by Central time. Again the hostess explained that our breakfast had been ordered at Oklahoma City and then canceled, because the plane was late. Lunch had not been ordered, and this was all they could obtain at short notice. I wasn't too hungry and felt to be thankful for this. But again I was thirsty. Oh, for some mountain water!

Flying over the Mississippi River was interesting, but seeing the many flooded farms on farther east was rather appalling. We were low enough at times to see buildings floating about. Our little hostess came and stood by us for a while, seemingly wanting to talk and to point out various places. I felt a sweet love of the Lord flowing toward her, and spoke of how very grateful we were for the plane and for their kindness to us. She then began to tell me in detail how very difficult it had been for them to get a plane and assemble a crew at such short notice. When I heard the sacrifices made by each one, and all the difficulties involved, again I marveled at what the Lord had done in answer to prayer. She said that to her knowledge no other group had ever been able to charter a plane in such a few hours' time. "It is practically impossible," she said. Blessed be our God of the impossible! "Of course," she said, "we are sorry to have to use this old plane, but it is the only one we had available." At just about this time, "This old plane" began to toss around and make even more noise. I wished it weren't quite so old! "Fasten seatbelts," flashed on, and our friendly pilot, in the most cheerful voice imaginable, told us that there just *might* be some rough weather ahead, hence we should remain in our seats. To my dismay I saw very black clouds rushing in on us from the south, and soon they were thick around us, and we could see nothing. We began to hit air pockets and to experience the inevitable sudden drops. Yet it appeared that many of the passengers were completely relaxed, and some were even asleep, including Dwight. I wished I were! Our flying kept getting rougher. And gradually the plane grew completely silent, except for the noise of the motors. It almost seemed that we were on a ghost ride—no one spoke, and no one could move much, with seatbelts fastened. (Later someone said he thought most everyone was praying by this time, for it was really turbulent.) I began to call upon the Lord in earnest, for I felt a strong temptation to become afraid or to cry. By now it was growing darker and I began to count how many hours I had been awake. It was now about thirty-four. I could not remember ever having been without sleep for so long a period, except during childbirth. And this made me feel that I was in travail—and that "things will be much worse before they will be any better," a favorite epigram for women in confinement. There is no turning back or aside in childbirth—there is only one way, and that is *through*. And this seemed to depict our present spiritual state—for we are all in the process of bringing forth the Body of Christ and the creation into glorious new birth. Soon the plane was tossing and rocking more violently and lightnings began to flash. It was raining very hard. It took a desperate interior exercise to keep me sitting quietly and outwardly relaxed in my tight seat. My mind was racing and I felt that at any moment I would have to rise and flee. But on a plane there is no way of escape except by parachute, and this way is only for the trained jumpers. In the meantime I must just sit it out. We later learned that the pilot had been instructed by the Tour Company to fly *through*—and not around or above—the storm, in order to get us to New York as early as possible. This surely tied in with my dealing about going through—and not seeking a way of escape.

I quoted over and over the line from the song the Spirit had deeply impressed on me, "My heart shall know no anxious fear." I knew it was like a command. Fear is deadly, and is God-denying. It is like a poison or a drug, paralyzing the strongest in its demonic power. Fear and faith cannot dwell together, "for whatsoever is not of faith is of sin." I had to resist it with all the power I could summon. The other word that came to me was a Scripture which had been much quickened to me for the trip: "Do *everything* in *dependence* upon the Lord Jesus, giving thanks unto God the Father by Him." I was to depend upon Jesus in everything I said and did, and not act independently in even the smallest things. At this present moment then, I must depend upon Him for grace and faith. And I must also depend upon Him to safely fly and land this plane. Having learned that the pilot had been called back to duty before he had rested, and had now been flying continuously for far more hours than the law allows, I felt a little dubious about him. Yet we could do absolutely nothing to help the pilot. We could not even give advice or make suggestions. All we could do was remain quietly aboard. How very like this is our passage from earth to heaven with the Lord. It is only He who can safely transport us onward and bring us to a safe landing. Thinking of this, a song came bubbling up out of my quivering heart: *The Christ Will Our Pilot Be*. I used to teach it to the children when I conducted Daily Vacation Bible School—some thirty or more years ago. But all the words were still clear:

> The Christ will our pilot be,
> A wonderful guide is He,
> So we'll sail, sail, sail,
> Christ will our pilot be.

I seemed to visualize Jesus Christ sitting up there flying the ship, and I sang it, "So we'll safely fly." With Him at the controls, what need I fear? Gradually the storm lessened and there were no more lightning flashes. We had a few minutes respite from our seatbelts. Our pilot told us we should be landing in Idlewild (N.Y.) Airport in about an hour—and that it was raining there.

As the passengers began to stir, the chorus of complaining again began. It was now nearly eight o'clock New York time. Where was our dinner? And what would we do for a plane to Europe, since ours was to have left at seven, and would not wait? There was much agitation and speculation about what would happen when we finally got there. To say the least it was disturbing. At this point I felt that I had flown all I wanted to for the rest of my life. To get on another plane and take an even longer flight across the sea seemed to me a terrifying prospect. If only I could at least rest a little first! Others were voicing similar thoughts, and were saying that likely we'd be shipped over on some old crate that might land us all in the sea, etc. I was relieved when we were told to fasten our belts again for some more rough weather, for this quieted us down. That last hour was in some ways the worst. It seemed that we were plunging madly into the darkness with nothing ahead or behind except oblivion. The motors labored and groaned, and I had to fight a crazy fear that at any moment, they would stop altogether. I also recalled having read that landing in New York in rainy weather was a tricky and uncertain thing. They have so many planes coming and going that when weather conditions are bad, they sometimes have to "stack them up" above the field and keep them circling, sometimes for hours, until they can land them safely. The thought of circling in the darkness for hours was ghastly. It was now nearly nine o'clock, and we were told we would soon be coming in for a landing. Yet we flew on and on and on—seemingly endlessly. We could see no signs of any lights below, and I began to think we must be out over the sea, instead of over land. But of course only the clouds obscured "little old" New York. I can assure you that all the "bridal" and "honeymoon" aspects of my journey were likewise obscured. I looked at the lovely corsage, now somewhat wilted, and wondered how anything this nightmarish could have seemed so desirable to my dear sisters. I even had a fleeting temptation to feel sorry for myself and think, "I just wonder how they would like staying up all night, being cooped up in here all day without much to eat, and being tossed about in a storm as well." I was almost willing to change places with any of you who would volunteer to do so. Almost, but not quite! (I repelled such negative thoughts of course. But anyway, forgive me, sweet doves!) Yet, with Jesus as our Pilot, how could I doubt and be so fearful? I was truly in a travail by now, one moment praising and rejoicing in Him, the next moment feeling the pain and pressure of fear tormenting me. It was like labor pains—I was actually *laboring* to *believe*. And I knew something must happen soon—It did!

We began to descend quite rapidly, and misty lights shone out of the darkness. They increased and we gradually settled down for the runway contact. It seemed almost like a miracle to feel the good terra firma again under the wheels. I could not restrain my tears. "A wonderful Pilot He!" I turned to Dwight and said, "If I did what I wanted to do, I would get off and find a train and go home." He laughed and seemed surprised that I had not enjoyed the ride more. He had blissfully slept through nearly all of it! This almost irked me, but I had to laugh too. "Anyway," I said, "I wouldn't do all this flying for anybody in the world except Jesus."

It *had* seemed like a big dose of flying for my first venture. Usually the Lord more or less eases me into severe trials, giving me little tastes first to accustom me. But in this case I had to take a big gulp instead of a taste. I was so delighted to get off that plane that it was a real test to have to wait for everyone else ahead of us. Yet I left at the last a little reluctantly. Had not Jesus been our pilot, bless Him! I know He was. For our pilot, in telling us goodbye, closed with these sweet words, "And now I say, God bless you all, and may you have a most enjoyable visit in the Holy Land." Our little hostess echoed, as we bade farewell, "Goodbye, and may God bless you!"

THE ASCENSION

appy is the Bride the rain falls upon!" This old saying flashed through my mind as I stepped off the plane and felt a few caressing drops of soft spring rain. A blessed quickening and a sense of joy ran through my being, and I realized that the Holy Spirit had ministered to me at that instant in a special way—perhaps in answer to your prayers. Once again I felt bridal and like I was on a honeymoon! A smiling Pan American crewman was waiting at the bottom of the ramp, and he graciously greeted us and directed us to go immediately up to the main floor of spacious Idlewild Airport—one of the finest in the world. A canopy had been provided to protect us from the rain as we walked to the entrance. There another Pan Am man met us and hustled us along. "Your plane is waiting," he said, and told us where to get our boarding tickets. This was a thrilling word, for we had all wondered if it would still be there, since we had been due at seven and it was now about nine-thirty.

Once upstairs, Dwight lined up for the tickets, and I made a quick survey of the interior of the Airport. Having read about it in advance, I was eager to explore it. But no time for that! I did find a place to buy a stamp to mail a card I had written to Jody. I also located a telephone, for I felt impressed to call home. I circled around again to Dwight and he told me the exciting news—we were going to fly by jet, instead of by propeller plane! I could have shouted! This would mean that instead of taking around 14 hours to cross the Atlantic, we would make it in half the time. I rushed to the phone to convey this reassuring word to all of you. Knowing how much delay one sometimes encounters making long-distance calls, I surely looked to the Lord, and depended on Him, as the Scripture says, to put the call through. The ticket line was already breaking up—there wasn't a minute to be lost. I got the operator, but she started to say that there were no lines available. However, right in the middle of the sentence she cut herself off, put through the call, and I was talking to Marian as clearly as if we were in the same room. In fact, I sounded so close that Marian at first thought I was in our village when I said, "We are in Idlewild." It took her a moment to get over the shock—the thought of our having returned home, after all that had happened, was too much for her. I learned that Jody was doing very well, and that everything was going fine. And I quickly sent my love and said goodbye—feeling all of you very close at that moment. This would be my first time to leave the U.S.A.—except for an excursion or two over the border into Mexico in my childhood. And here it was so dark and raining I couldn't even see my native land as I departed! There would be no long, last, lingering look at the Lady of Liberty for me.

I hastened back to Dwight and found him talking with Dr. Hollander, the head of Compass Tours. Dwight was thanking him for having gotten this plane and now a jet. Others thronged around, asking questions. Only a few seemed thankful. Most of them were complaining about all they had gone through, and were telling him how upset they were. But Dr. Hollander held his peace and kept reassuring us that we would reach our hotels in time, and that we would have a fine trip, in spite of this initial trouble. But the complaining continued, and I felt again a sense of shame and dismay at the very carnal attitudes displayed. Soon we picked up our bags and went to our appointed gate. Our planeload of about 120 had been divided. More than half of us would leave at once, the others must wait for about a half hour. This caused more upsets and complaining. (Later we found that the second jet did not leave until the following morning, because of rain.) Our entry into the Pan Am DC-8 Jet was most regal. The welcome carpet was out, and over us was spread a nice canopy. (Instead of the old type "gangplank" the jets used a bellows-like portable "jetway.") Every few feet a gracious crewman or woman greeted us personally. Since we had all been given seat numbers, no one shoved and tried to run ahead to get the good seats. Like ladies and gentlemen we decorously made our way up the ramp. How huge this great bird, with its four-story tail, appeared to me! And the inside was truly luxurious, and much like the pictures I had seen on T.V. But of course this was the first class compartment. We went on into the Tourist Section— which occupied almost three-fourths of the ship. It too was very attractive and seemed spacious and gracious after our other crowded plane. Dwight and I had three seats instead of two—and could put our bags on one of them, giving us plenty of foot room and space to move about. At each seat were several things for our comfort—a little desk-like tray that let down for reading, eating or writing; special lights, ranging from dim to bright; an individual air-conditioner; booklets and other items for our personal use.

And the seats were much more roomy and relaxing. I put on my gay little slippers and wished I could take off my tight suit. But now I was growing very weary of it. A hostess brought us pillows and fluffy blankets, but I was too excited and wide awake to think of sleep. My little song about the trip was playing on my hidden "recorder":

> To the Holy Land I must go
> To walk where Jesus trod.
> And the time has come to rise up
> And follow the call of God.
>
> *cho.*
> Catch me up, dearest Lord Jesus,
> Up above the clouds with Thee.
> O, may this flight give Thee delight,
> As we soar in harmony.
> And then let me be Thy glory
> When in Jerusalem I stand,
> And may I magnify
> And lift Thee high
> Throughout all the Holy Land!

Suddenly I remembered that it was approaching midnight and that Ascension Day was at hand. At Easter time the Lord had spoken clearly to me about the forty days between it and Ascension Day and I counted them off carefully. It thrilled me to then discover that I would be flying to the Holy Land on the actual Biblical Ascension day. "Thou hast ascended on high," kept ringing in my heart. Now I sang inwardly the last verse of the song:

> Angels and saints shall attend
> Our pathway through the sky.
> And the wings of the dove shall bear me
> To the Father's breast on high.

About this time my sweet meditations were interrupted by the voice of our pilot. He welcomed us all and apologized to the other passengers for the delay caused by waiting for all of us. However he graciously reassured us that since the radio system had needed repairing anyway, a delay had been inevitable. He told us that our destination was to be Frankfurt, Germany—a distance of about 4,000 miles. What a surprise! We hadn't expected to go anywhere near that part of Europe. We would cross five time-zones enroute, hence would lose five hours. Prospective time of flight—6 hours; average speed—around 700 miles per hour—almost as fast as the speed of sound! Our elevation, he said, would be from 37,000 to 39,000 ft. This would be by far the fastest and highest I had ever traveled in the flesh! Our Captain also electrified the atmosphere by telling us that cocktails would be served as soon as we were airborne. A loud clamor arose and cries of "We're starved," "When do we eat?" "We don't want cocktails," and "We haven't eaten all day," mingled throughout the ship. These seemed to offend some of the other passengers, and I felt ashamed of our childishness. The unfailing courtesy of the crew made our own rudeness seem the more flagrant.

The confusion died down when the Captain courteously announced that since some of the passengers had not had dinner, they would be served promptly. I suddenly felt very hungry. It *had* been a skimpy day, and a long time between "snacks!" The next words of the Captain took my appetite all away. We were now ready for takeoff, but one thing remained: one of our hostesses would demonstrate the proper use of our life-vests. Our life-vests! Sure enough, they were in a pocket of the seat, clearly labeled. And on them was detailed instruction about their use when abandoning ship. Our smiling, dimpled hostess was gaily putting one on and telling us just how to unfold them and get into them, and warning us not to inflate them until we were outside the plane. The very thoughts of it deflated me to minus zero! Just to get out of the plane would be more than I could manage. (For a few moments this is just what I wanted to do.) But then the engines began to purr and it was too late to do anything but sit tight and go through with it. I knew that Jet takeoffs were extremely noisy, and I dreaded the shrill siren-like almost unearthly "screams" I had heard standing and watching them go. On the plane however, the noise was not nearly so penetrating. And soon we fairly zoomed off in the air and rapidly soared.

It was all so smooth and comparatively quiet that I was delighted. And this continued all during our flight. Truly it seemed like a "magic carpet" or a great bird. After being on a propeller plane, this was indeed a "heavenly flight." There was no awareness of speed at all—we seemed to be floating. And there seemed to be a celestial atmosphere all around. By the time we reached New Foundland it was clear enough to see the lights below. And soon the northern stars shone out with great brilliance. Our window faced toward them, and I joyfully greeted the north-star and Cassiopeia—The Queen's Throne—which was right beside us. There was a feeling of flying *into* and *with* the stars—a most exciting and rapturous thing! But soon very earthly aromas were wafted around us. Dinner was about to be served.

Those of you who have flown know that the serving of a meal on a plane takes a lot of time, depending, of course, on the number of the passengers and of the crew. We had two hostesses and four helpers, I believe, just for our compartment. But they were quite a long time reaching us. There is a tendency for the passengers to get very restless about mealtime, and to begin to jump up and down, or ring for pillows, blankets, a drink or some other attention. All this delays the meal by crowding the aisles and distracting the hostesses. During this interval the pilot came back into our section and paused at each seat for a short chat. (There were two co-pilots and three engineers aboard.) He answered various questions and assured us that jet flying is the safest way yet found. He was charming—"Christ will our Pilot be." It was eleven-thirty before our dinner was placed before us. But what a meal! A roast half of a duck, wild rice, beautiful vegetables and salad, and special little appetizers. But no cocktails! They were never mentioned again. (This was one of the two things that made us a curiosity everywhere we went—that we did not drink, and the other was that there was no smoking among us. For this we truly were grateful. If only there had been no fussing and complaining, it could have been heavenly.) By the time we all had finished our delectable dessert and had seconds of coffee or tea—it was about one o'clock. The lights were dimmed, and everyone let down their seats and prepared to sleep. I tried, but it was no use. I was absolutely wide awake. It seemed incredible that by this time I had been without sleep for forty hours. I felt very much as I did that night in Sequoia—of which I have written and you have read—when I was taken up in the Spirit and soared with the Lord in a most wonderful way. This night was similar, yet so very different. I seemed semi-rapt, and was in a constant state of interior song or music, with praise and love flowing toward the Lord.

The plane grew very quiet and this helped me to be detached from our surroundings. I was also a little spellbound by an ethereal misty light which I could see from my window—it seemed to surround the plane—and it appeared truly supernatural to me. It kept increasing, but I finally realized that it was the approach of the dawn—at two o'clock in the morning, New York time! (Though by now it was about four, since we had lost two more hours.) Oh what a dawn! There will never be another night and morning like that as long as I live—it seemed almost like Creation's first dawn, and the appearance of this light in the clouds beneath us was comparable with a picture I have seen of God separating the light from the darkness at Creation's first misty morning. I tried to enjoy and appreciate it to the full—and to glorify the Lord in it too (in an interior manner.) I felt some of the rapture songs: "At the Breaking of the Day," "There's a New Day Dawning," and others. And a beautiful Scripture, long cherished, was vibrating within: "If I ascend up into heaven, Thou art there . . . If I take the wings of the morning, and dwell in the uttermost parts of the sea; even there shall Thy hand lead me, and Thy right hand shall hold me. If I say, surely the darkness shall cover me (as it had in the early part of this night); even the night shall be light about me." The light was steadily increasing. And at exactly 2:30 the sun itself arose and was visible through the clouds, a pale rose-colored orb in the east. It was fantastic—but true! Oh night of nights! Oh dawn of dawns!

About this time there was a stirring aboard. The rest-room procession had been quiet for about an hour. It started again. We learned that electric shavers were available for the men, so a lineup soon formed. Others said it was breakfast time, and began to murmur about being hungry. And soon the scent of coffee began to permeate the atmosphere. (From the heavenly to the mundane is often just a breath or two!) Our breakfast arrived at about five o'clock, New York time. It was now nearing nine o'clock by actual time. And our pilot began to converse with us again by loud speaker. We would soon be over Ireland, he told us, and if the clouds did not hide it, we would have a fine view. We gradually lost altitude and soon caught our first sight of LAND! It was hard for me to realize how long our ancestors had to sail in crossing the Atlantic before they sighted it. Old Ireland appeared—green and jewel-like, shining out in the morning sunlight. The emerald isle indeed! But it was hyacinth too! In that light all the water appeared not blue, but hyacinth. This was a scene of unforgettable beauty and splendor. The Irish sea was like a great Jacinth, set in the midst of emeralds. I thought of our Jacinth mountain, and, at the same time, of a song I loved to sing as a child: "A Little Bit of Heaven." It referred to Ireland, of course. I also thought of Norma, who had sponsored this land as one of her nations. By this time we were over England, near London. But, alas, the clouds closed in again. We saw only a misty view of London and the English Channel. And Belgium modestly hid entirely, as we passed over it. Since these three nations were all especially of concern to Norma, I seemed to commune with her as I blessed them. Soon we were over Germany, and this brought thoughts of Mary. I was praying for each nation as we traveled, and felt that I was dropping "love-bombs" upon them, or that some secret weapon was raining down rays of life and love from the plane.

The beautiful and thrilling ride was nearly over. Our pilot told us that during the flight the temperature around us had reached minus 68 degrees Fahrenheit! But we hadn't even been chilly! It was raining slightly in Germany too—and everything looked bright and green. The red-tile roofs of the houses soon grew closer—only then did I realize we were descending. We glided in with the greatest of ease, like the man on the flying trapeze. We were nearly 7,000 miles from home! But I felt very much a part of Germany, and wished we might linger there and walk among its people.

As we alighted, we saw very close to us a smaller plane, a "Comet" Turbo-Jet— and on it was the insignia of the Cedars of Lebanon! My heart leaped when I saw it. It made the Holy Land seem very near. I had felt an urge to mail a card to Mary from her land, however I realized that I had no German money with which to purchase a stamp. I soon discovered that we could not even leave the field since we had no entry tickets. But a kindly Pan American Hostess offered to supply the stamp and send it on its way. We were almost rushed to the "Cedar Jet," as it was called. The crew stood smartly at attention and pretended not to notice how some pushed others aside in an attempt to get the best seats. However polite these men and women were! They appeared to be Lebanese. And I soon discovered that they spoke German, French or Arabic better than English. I truly felt the presence of the King, and rejoiced that this would be the last major step toward reaching His Land.

A delightful German-style luncheon was served to us soon after takeoff. And it was certainly relished by all aboard. It had seemed a long time since breakfast. I felt as though I was feasting with the Beloved as I partook of this. And I kept looking out and blessing Germany, then Austria (Elda's land), then Yugoslavia (Gertrude's), and finally, Greece, (Jennie's), as we flew over. Our pilot said the clouds were clearing and he would try to give us a good view of Athens. For the most part we had flown over green fields, woods, houses and small villages—no large cities. As we turned and came down fairly low over Athens, it was positively breath-taking, and much like Cinerama. I was truly thankful to the Lord for clearing the sky. There were just enough white clouds to contrast with the heart-breaking blue of the Aegean Sea. The creamy white city of Athens appeared most regal and glorious as we circled it. Then on we went, over several other little islands, heading toward Cyprus. We saw Crete at a distance. There was a lot of speculating about what we were seeing, and some displayed quite a little geographical ignorance in their pronouncements. One that struck me as the funniest was that of a solemn brother, who pointed out "Sicily" to us—it was actually far, far from our course. Since the pilot had promised us also a complete view of Cyprus, I was most thankful that the clouds parted again, and we dropped much lower as we neared it. Majella was with me at this point. I recalled how much she loved this dear island which she has sponsored, and how eager she had been for me to visit it. I watched carefully as we came down over its western portion, and passed along its beautiful coastline, crossing over near the east. The pilot pointed out Nicosia, the capital, and it appeared very beautiful. It was to this island that St. Paul and Barnabas journeyed on that first famous Missionary Journey, I suddenly recalled, and now four-fifths of the Cypriots are of Greek Orthodox faith, it is said.

After leaving it behind there was only the Mediterranean for a long ways. How blue it looked too, a little deeper and darker than the Aegean. "The sea in the heart of the nations," it has been called. How can I tell you what I felt—no words are adequate. The ancient world was at hand and I was gazing on the only real sea my Lord and most of His early followers knew. Most all the Bible references to the Sea refer to this one. (Since both Galilee and the Dead Sea are more properly lakes.) I loved it! Yet I was eager to see land, for when it appeared we would be actually in the Holy Land in Lebanon.

We had lost some more time enroute, so it was around five o'clock when we crossed the beautiful coastline and came down for a very smooth landing in Beirut. Of course Eddy was with me as I landed, this being one of her countries; and Marian too, for she had prayed earnestly that I go there—had felt much about it. I was indeed full of joy when I saw the dear faces of the native Lebanese, watching our arrival with friendly interest. I was afraid of my Arabic pronunciation, so tried French, and it drew a quick response. I just spoke to everyone we passed. And what a bowing, smiling and enthusiastic response was displayed by them. We were quickly corralled in a large lounge, and asked to wait for a few moments. Since we had no visas for Lebanon, we could not step out of the station. Indeed, I am sure we were not supposed to leave the room. But I felt to send more postcards, so I went over to a counter to buy some. Then I realized that I had no Lebanon money, nor could I understand what the boy kept telling me. At length we made a bargain—one postcard for one American dime. But now I had no stamp!

Also I was very thirsty. How I longed for a drink, but again, I had no money. If I gave them a dollar and they took it, I would get my change in their currency, which I would have no time to spend. But this I decided to do. About this time they began to serve all of us cold drinks free, giving us a choice of several kinds, most of them American. How delightful! I never did know just who paid for them, but we all enjoyed them. It was hot!

I was so eager to get a stamp that I left the lounge again and almost started running toward some stairs, hoping to find someone to ask about the Post Office. Suddenly I landed on both knees—I had overlooked one little step down! Almost immediately three charming Lebanese men were at my side, tenderly lifting me up and making solicitous inquiries, mostly in a language I didn't understand. I assured them I was all right. But I felt like a fool. However I asked them in French where the Post Office was. And they understood me! But the problem was that I didn't understand what they replied! I did gather that I was to go up. So up I went, inquiring two or three times, and each time getting an answer I couldn't quite make out. But I finally located the *"Bureau du Poste"* and found a really delightful Postmaster in charge. He was full of greetings and questions, half in English, half in French. Where had we come from? How long would we stay? etc. By this time I was really concerned about the time and getting back to the group. Not even Dwight knew where I was. So I tried my best to buy an air mail stamp. But he couldn't see why I needed an air mail stamp for a post card. (Almost no one abroad could understand this!) He argued! I persisted! And then a man came up to him and they began a lively conversation which was more Greek than French to me. It seemed impossible to get that stamp. Finally he gave me his attention again. And just at this point one of the stewardesses came running toward me. She was not so gracious as others had been. "Madamé! Madamé!" she cried out in real agitation, "Come at once, Madamé, the plane is leaving." This electrified me into action, and I left my American quarter and my card and did a fast sprint after the stewardess. It was true! They were all on the plane and Dwight was indeed agitated. I gathered that I was not supposed to have gone wandering around "out of bounds." The other passengers were not pleased at being kept waiting, and I learned thus early in the Tour that he who keeps the others waiting is a "baa! baa! blacksheep" indeed.

My beautiful Ascension Day had come to an end. It was now growing dark, and we were again crowded into an old propeller plane. Instead of being headed toward Jerusalem, we were enroute to Amman, Jordan. And to make things even more disappointing, it was too dark for us to catch a glimpse of beautiful Mt. Hermon, the highest and most conspicuously beautiful mountain in the Holy Land, which has always been of special interest to me. Its three summits, situated like the angles of a triangle, seem to me to depict the Holy Trinity. This was the Mount Sion of Deut. 4:48. From this mountain David took the name and applied it to his chosen mountain in Jerusalem. To be in Lebanon and not to see Mt. Hermon was indeed painful. And of course there are also the famous Cedars of Lebanon and the scenic splendor which is referred to as "the Alps of the Holy Land." Of great interest too are the Phoenician Sea cities, the Crusader Castles and the Roman ruins at Baalbek.

Things were not going very well with me: I had visited Lebanon, and yet could not actually see any of it except the coastline and portions of Beirut; I had fallen down; I had kept the whole plane waiting; and now, to my dismay, I discovered that I had also lost my precious glasses! As in the case of previous "Ascension" times, I had descended suddenly with quite a thud! After having flown so high, reaching the highest altitude I had ever experienced, I was now on my way to the lowest place on earth—the Jordan River near the Dead Sea. From the heights to the depths! From the depths to the heights again? Oh may it be so!

THE DESCENDER

N ow we were bound for Jordan, praise God! And it was an awesome thought that within the span of twenty-four hours I could ascend so high and descend so low. The Word in Ephesians came to mind: "When He ascended up on high, He led captivity captive and gave gifts unto men. (Now that He ascended, what is it but that He also descended first into the lower parts of the earth? *He that descended is the same also that ascended* up far above all heavens, that He might fill all things.)" I had followed Him in His ascending. Now I must be equally willing to follow Him and His descending. And I recalled then that Jordan means, "The Descender." This name given now to the land belonged first to the river which flows through it—the most famous and beloved river in the world. During its journey from its source to the Dead Sea it descends 3,000 feet, winding back and forth in a serpentine fashion. The alluvial valley through which it passes is a part of the strange Ghor (gash), which is a great rift in the earth's surface, the like of which is found nowhere else on our planet. (I have wondered if this did not in some way symbolize the riven side of our Lord, from which flows the river of salvation.) This valley is known as "the cellar of the world." It reaches its lowest point at the Dead Sea, 1,300' below sea level. The rift itself is even deeper in the basin of the Dead Sea. Yet the Jordan river finds its source in beautiful Mt. Hermon, 9,000' above sea level!

It has taken a paragraph to summarize the impression I had as our plane took off—so very near Mount Herman!—and began its flight toward Amman, the capital of Jordan. From thence, we were told, we would be taken via bus to Jerusalem. This would mean passing through the Jordan valley. We would be going into the depths in a realistic way. And already I seemed to have arrived there in my spirits. I was indeed deflated and tempted to become depressed. I felt strangely disturbed and humiliated. And soon my condition worsened. All around us was the sound of murmuring and complaining. Everyone agreed that we should have been taken to Jerusalem directly, rather than to Amman—a place seemingly unfamiliar to most of the passengers, and thought of by them as "way over in the wilderness of Jordan." Complaining gave place to indignation in several. "When will dinner be served?" was also repeated over and over. Our hostess was very calm and obliging in answering questions. Our flight would not be long, she said, so no meal could be served, but somewhere enroute by bus from Amman to Jerusalem, we would be given our dinner. The trip would not take long, she said, since the distance is only about 60 miles. This disturbed me as well as others. I had read about that road from Amman to Jerusalem! Unless some drastic changes had been made, it would take us a long time to get over it. I too was eager to get to Jerusalem. But I, and apparently I alone, understood why we were not being flown there directly. Since I was thus informed, it seemed to be my duty to tell others. So I joined into the general conversation and, as soon as I had the attention of some of the ministers, I announced that the reason why we could not fly into Jerusalem so late in the day was because it was Friday, and therefore the beginning of the Sabbath. I patiently explained that it begins about 3.00 p.m. and that all traffic stops then until Saturday night at sundown. No one seemed the least impressed by what I said, or even to pay any attention. They went right on discussing it. So I retreated to my own thoughts. It must have been at least five minutes before I suddenly saw what a ridiculous thing I had said. It was indeed fortunate for me that no one *had* listened to me. Of all the geographical ignorance displayed, mine was the most gross! The Jerusalem we were going to is in Jordan, and in Jordan the Sabbath is completely lost—since not one Jew remains! (Later we learned that the reason we could not be flown into Jerusalem was because they have no lights for night landings.) I was really humiliated by this display on my part, and began to seek the Lord earnestly. All along the way I had felt painfully ignorant about the geographical and historical aspects of the land we had flown over—including our own country. Some way it all looked very different than it had on maps and globes. Yet I had felt I would have an understanding of the Holy Land, having studied it more thoroughly. Instead I discovered that I was confused even about Jerusalem, and was thinking of Israel while heading for Jordan. Soon the Scripture came to mind, "Do everything in dependence upon the Lord Jesus." As I considered all the mix-ups I had been in, I realized that I had not consistently sought the Lord's guidance in my movements, nor had I depended on Him when I tried to buy the postcard. Instead I was relying on my French and English. I had rushed to the Post Office in my own spirit, falling down enroute, (as we often do when we hurry along in our own will.) I had tried to buy a stamp in my own efforts, not really looking to the Lord. I had lost my glasses, delayed others and worried my husband. Then, safe on the plane, I had spoken about Jerusalem in my own spirit too—not waiting for a leading of the Lord, and perhaps feeling slightly superior in my heart to those who seemed so ignorant about the land. All in all, I had played the fool. And it was because I had not really *sought the Lord in each little thing.* Although for many years He has trained me and exercised me along this line, I still often fall short. (I realize that none of us can be consciously guided by the Lord in every single move we make, no matter how carefully we seek such guidance. However, we have learned that there is a sweet "reining and constraining" of our thoughts, words and activities, when we seek and depend upon the Lord in every little—as well as big—thing. I remember a dear old saint telling me that only the *truly dependent* soul learns the secret of having the Lord's assistance and blessing in every detail of human life.)

Usually, when I am on a trip with the Lord, I am extra careful about even little things. Often *much hinges on little.* There, in the plane, I saw how by acting in *independence* instead of in *dependence*, I had failed, fallen, suffered loss and become depressed and confused. I truly cried out to the Lord to forgive me and to let all this be a lesson to me, and to help me to be very careful for the rest of the trip. I felt that I had gotten off to a bad start.

Again it has taken longer to describe my interior dealing than for it to take place. I began trying to praise again, within my heart, and attempted to calm the anxiety I felt. The plane was so very noisy and old that it made me nervous. And I was aching from head to foot, just from being weary and cramped. I noted that I had not injured myself in the fall at all, not even my stockings were torn. But after wearing clothes for two days and two nights, they do become most binding and uncomfortable. I began to wonder how long it would be before I slept, and just how long one can survive without sleep. It seemed that by now I was in an odd mental state from being awake so long. I really don't know how to describe it, but it was like being in an unpleasant dream. I was also beginning to feel very shaky. I had dreaded nerve exhaustion more than anything—having gone through times of nerve-collapse in the past and thus knowing some of its torments. How I wish I had gotten that extra rest before leaving! I repented of this disobedience too. And how I wish I had brought an extra pair of glasses or at least my prescription, as all the travel books had said to do. I had thought to economize by not doing so. But how I regretted it. To read, take notes, dress, pack—and do many other things, I need glasses. I knew the Lord could deliver me or wonderfully help. Yet I realized that I had failed to make the necessary preparation for my eyes.

But there was no use dwelling on these failures, once I had confessed them. So I worshiped and praised the Lord interiorly. About that time our Hostess said, "Look, now, and you will see the lights of Damascus." Damascus! I had not dreamed I would catch a glimpse of it, though I so longed to go there. Suddenly St. Paul was very near. Why, for all I knew, we might be flying now over the very road he traveled! We might even be near the very place where Jesus appeared! Since we were flying very low, we had a good view of the city, though I still don't see just how we got so close to it. Had we flown *over* those high Lebanon mountains? We veered off southward and flew on toward Amman. Few lights appeared beneath us, for Jordan is one of poorest of the poor among the nations, and has few cities. But Amman itself is teeming with activity, and is one of the fastest growing cities in the world. I tried to remember what I had learned in studying about it—but was too weary to recall, as I do now, that its population is more than 200,000. In 1936 there were only 20,000 living there, in a mud-walled village with dusty streets. Yet now its streets are filled with American cars. And all about are new buildings. East and West meet here in a strange medley. As our plane flew over it and circled for a landing, it seemed like a little town—in comparison with our giant cities. It was hard for me to realize that this was the actual capital of Jordan, and that here the King lived, and everything had its center. (I am sure we are all glad that the Jordanians deemed it unseemly to make Jerusalem a commercial or governmental center of their country. They are trying to preserve it more as a Holy City. Their name for it—*El Kuds*—means "The Holy." Much of the business that once took place there is now centered in Amman.) It is also a camel-trading center. Western stores and Oriental shops are side by side.

Donkeys and camels vie with buses and taxis in the streets. But none of this was visible as we touched down. The airport was rather new, but ever small. I was glad we found the runway—which was not very well lighted!

As we taxied to a stop by the main entrance, a bright light turned into one of the most beautiful stars I have ever seen. I am sure the haze on our window may have caused this phenomenon—but in any case I was dazzled and thrilled. It was blue and gold and white—and appeared to me like an actual star that had followed our descent out of the heights and had come to rest in dear little Jordan. A modern star of Bethlehem to me! Following, it, I walked off the plane—amazed that to my eyes it still looked like a star. Again I was delightfully quickened. Love fairly throbbed in my heart—love for Jesus and for these dear Jordanians, of whom we had read and for whom we had prayed. I wanted to kiss every one of them! On a balcony above the entrance, a little group in native dress was intently watching us. When I looked up and waved, they all waved back, and the children shouted, "Halo!"—their particular form of "hello"—a word we were to hear on every hand a hundred times a day in Jordan and Egypt.

Once inside the very hot airport lounge, it was most disappointing to see it filled not with Arabs but with Americans! And it was even worse to find that they were tired, thirsty, hungry and upset, just like our own crowd. I squeezed around to a chair adjoining one circle of prosperous looking tourists, and heard their Tour Leader explaining to them that we were those Pentecostal fanatics who were going to take the rooms in Jerusalem they had been occupying. I also heard him say that probably we would all be going to put on robes and be baptized in the Jordan, as some of the Greeks do. At that very moment, in fact, a Grecian group wearing white robes passed through the door. They were obviously on their way to the Jordan— perhaps to be baptized at dawn. As soon as this man left, I felt led to speak to these women, who were from Covina, California. How surprising! They began to tell me about what they had suffered that day. At noon, it seems, they had been hustled out of their hotel rooms—so that the rooms could be prepared for that "Pentecostal Sect" that were coming. They had been left sitting for hours in the hotel. Then, by bus, they had been driven to the Airport. It had taken them two and one half hours just to get through customs. Then, on top of this, they found that their plane had been "purloined" (probably by us) and was very late. So, of course, they had not been given dinner, and this airport had little to offer in the way of food. About this time the Spirit moved on me to identify myself and our group as Pentecostal—and to tell them about our night-long wait in Los Angeles. The Spirit began to laugh in me, and I seemed to overflow with love toward them.

They stopped smoking and stared at me. Then one said, "Well if that is what it is to be Pentecostal, it must be great. Imagine being able to laugh about all that." I felt to speak then about Jesus, the meaning of the land, and how wonderful it was to be there. I could see that their tour was evidently only for pleasure—for they listened intently, and seemed impressed. One had tears in her eyes when I had to go.

I did not seem to mind not having food, but I was truly thirsty. I thought that surely there must be a drinking fountain *somewhere*. So I went in search. I didn't find one, but I did find some Jordanians, a little cluster, watching us keenly. I said, "Hello!" And they "haloed" me warmly. I struggled to think of even one good Arabic word, and came up with *"Deekshee mezean!"* (That is nice!) I pointed to one man's Arab head-dress as I said this, for I was trying to say, *"Dey eh dey?"* (What is this?) In any case, he understood me, and told me it was a *kuffiyeh* (ka-FEE-ah.) (This sheik-like headcovering is used widely throughout Jordan and also in Egypt.) Over and over my new friend said the word, and of course I repeated it after him. One man who could speak English welcomed me for all of them. He was delighted that I knew a few words in their language. Thus encouraged, I tried a few more. I said most of them wrong, but they understood and corrected me. I felt at once a warmth and a child-like friendliness in them that was most pleasing. These people were frank and sociable. Their admiration for America was evident from our first contact with them. When asked why I had come, I told them that Jordan was the most precious of all countries to us, because of our love for Jesus Christ who was born there, ministered there, was crucified and resurrected there, and would return there some day. I spoke of His love and how it possessed our hearts. They listened with respect while this was interpreted by the one who spoke English. Then he smiled and said, "You Christian, we Moslem. But God is over all and we are brothers." As I stood looking into their dear faces I realized how very much it would mean to Majella (who sponsored Jordan), and to the others of our Company, to be standing in the land, having actual contact with native Jordanians. And I blessed them before I turned to mingle with our own people. My anointing to praise the Lord and talk about Him continued. I felt actually giddy with His joy! I knew all this was from the Spirit in answer to prayer—and I was most grateful.

Our stay at the airport was not too long, in view of the time it takes to go through passports and customs. They showed us special courtesy, and soon we were led out to the waiting buses. I had a little shock when I saw how small and antiquated they were—oh for a Greyhound about now! We were really packed into them, with all our hand luggage, cameras etc. And even the little extra aisle seats were put down. This meant that everyone was really uncomfortable. And since it was very warm and we were very weary, there was quite a bit of fussing. Some tried to joke about it. But the atmosphere was tense. I was dismayed when the men began to make fun of the bus and the driver. They did look queer, I had to admit. But it is a very important tourist rule that one does not belittle or make fun of anything in another country. The underprivileged nations feel it keenly when we laugh at their poverty and backwardness. They do not understand that Americans complain and joke about things just as much at home as abroad. How much I wanted to silence the men on this night. But I was resolved not to speak again without a real leading. As we made our way through Amman's narrow, crowded streets, the men kept pointing out things that amused them. I was near to tears by now, for it seemed very difficult to feel the presence of the Lord. And as I looked upon the noisy crowds and garishly lighted buildings, I had to strain to remember that this was once the biblical city of Rabbath Ammon, the chief city of the Ammonites, and that David had made conquest of it around 975 B.C. I vaguely recalled that much earlier than this Jacob had wrestled with an angel at the Brook Jabok not far from here, and that during the time of Christ this city was a part of Decapolis—and its Greco-Roman name was Philadelphia. Whether Jesus visited it or not is not positively known, but many Roman ruins remain unto this day, proving it to be a city of importance and beauty. Bouncing through it on this night, the past seemed almost incredible.

Soon we had left the city behind and were obviously out in the country. The main question now was, "Where do we get our dinner?" It was after nine o'clock and everyone was very hungry, since lunch had been seven hours or so ago. Someone said that we were to be given dinner along the way. What a bright prospect! The sooner the better! Having read about how bad the food might be in Jordan, I was not too enthused. But still anything was better than nothing. Sure enough, before long our buses drew up and stopped at some sort of Inn or Hotel. I remembered that I had read in Dr. Field's book that there was an excellent Inn about eight miles out of Amman. I began to picture a really delightful Arab dinner. A clamor arose in the bus. The problem was, how could we possibly get out and later back in the bus, so tightly were we packed. But this dread, like so many we were to encounter along the way, was groundless. There would be no need to leave the bus, for just about now a smiling Arab jumped on and began passing out paper sacks. A stunned silence fell upon us, broken only by the crunching sound of paper bags. No one, absolutely no one, could really grasp what was happening for a few moments. A paper sack *lunch* is always a remote possibility on a tour. But a paper sack *dinner* especially when one has been starved for hours—seemed just unbelievable! Several of the men sat in frozen positions, "holding the bag" literally, and staring straight ahead. By the time we had recovered enough to open the bags and talk, the bus was again merrily on its way, jogging, bumping and swerving along the road toward ancient Jericho. By now, of course, you are very curious about what we found in those memorable paper bags. You will hardly believe it—nor could we—but it is true: there was one sandwich of very dry Arab bread and a slice of cheese—nothing else on it. And there was one small orange. This was our "dinner." Righteous-American-indignation flared. Our one cheese sandwich enroute to New York had become by now a standing joke. Here was another to add to it—a double joke! But it wasn't funny that night. There were actual howls for coffee, for meat, for something hot. And almost everyone aboard was going to take this up with the Tour people. If this was a sample of what our Tour meals were to be—then certainly a mutiny was in order. One or two suggested suing Compass Tours for all we had suffered! Only a few tried to turn it off as a joke. And about this time our Arab lunch provider spoke up. (I wasn't even aware that someway he had crowded on and remained with us.) He explained that since our plane was so late, there was no time to stop for dinner. Our hotel in Jerusalem was waiting. We must go on and eat as we traveled. This, of course, re-opened the issue of all the things that had happened enroute to make the plane late. Ah me! What a turmoil it was. Yet in all that I felt a sense of peace, and a verse kept fluttering in my heart about the early Christians breaking bread from house to house and eating with "gladness and singleness of heart, praising God." It seemed difficult to realize that we were the spiritual descendants of that original Pentecostal church, and that we were on a "holy pilgrimage" to Jerusalem to keep the Feast again.

I actually enjoyed the bread and cheese, and the little rather sour orange. And then I suddenly grew very sleepy. The voices around me began to fade, and lo! I was asleep—for the first time since Wednesday morning at 6 A.M. (It was now after 10 P.M. on Friday night.) We were descending rapidly and the atmospheric pressure was increasing. But as soon as I was asleep we were *flying* again. Oh no! Yes we were, and on a bus! It was really very amusing in one way, for I could *see* where we were going, and the little barren hills were visible all around. We seemed to be only a few hundred feet above the ground. Then I suddenly thought, "Why this bus *can't* fly, it wasn't made to. We're going to crash!" Then I woke up, so very thankful to be safe on the ground. Safe? Well, one rather wondered! We were told that the present road from Amman to Jerusalem is really a fine one, in comparison to the old one. But at times it seemed precarious, particularly when other vehicles approached us and had to pass. Even some of the men were jumpy about it. But in comparison with flying, it surely seemed safe to me. So I went back to sleep, only to have the same dream over and over and each time to waken again startled and yet thankful. I suppose I did get a few minutes rest, but I think something must have happened of a supernatural nature too, for suddenly I was wide awake again, and quickened in a special way. I heard someone say that Jericho was off to the left and I realized that we were actually turning now toward Jerusalem. I knew that somewhere along the way we had crossed the Jordan by the Allenby Bridge and were now on the ascent again, going up—oh wonder of wonders!—to Jerusalem. The dream of a lifetime was about to be fulfilled on this dark-of-the-moon, Arabian night. And here I had been drowsing away, half unaware of the significance of it! We were making very sharp turns by now, and the bus was reeling from side to side. By its headlights we could catch glimpses of rocks, caves and very rough terrain. This was the road where robbers had lain in wait for travelers—the very road about which Jesus had spoken the parable of the Good Samaritan. Perhaps soon we would be catching a sight of the ruins of the Good Samaritan Inn, of which I had read. Was it possible that we were going to continue on up this road without a word of prayer, praise or Scripture? I quietly began to worship the Lord and praise Him in an interior way. But soon I felt the Spirit move on me to praise and sing outwardly. We have all learned to shrink from making a needless display of our worship before others. But there are times when the Spirit wants us to praise openly. I felt free to do so because others around me were loudly talking. As I sang the Spirit moved with increasing strength. I had a most marvelous communion with the Lord and with the precious saints of the past. I seemed to be in procession with them, going up with high praise, singing the songs of Zion, into the city. It seemed to be in some future time, rather than now. We, the ransomed of the Lord, were returning with singing and everlasting joy. (Isa. 35) I was almost rapt with worship and high and holy song. Yet it was all very soft and melodious outwardly. Then the song changed and was turned to weeping. The grief of the Spirit possessed me, and He spoke in tongues through me about His grief in the lack of the overcoming life among His people, the lack of fervent love, praise and joy. He also spoke of the coming Convention in Jerusalem and how He had purposed this great meeting with much desire to gather us together for a mighty moving of His Spirit. He wept and made it clear to me that it would fall short of His intended glory. It was oh so plain now that everything that had happened along the way had been designed to bring forth in us faith, praise and a closer communion with Him and with one another. I saw in a flash the great value of the years of His unfailing training and discipline in my own life and in our midst. Without all this I too would have fallen into complaining and doubting. Oh this wonderful life of praise! This wonderful, victorious high-way of turning *everything* that happens to us into an opportunity to manifest the grace and glory of God!

At some point, during this time, we did pass The Good Samaritan Inn—now used by police—but I was too much in the Spirit to give it more than a glance. As the Spirit gradually began to lift, I realized that we were climbing rapidly and must be nearing Jerusalem. How I longed for our first glimpse of the city! I felt I must continue to sing until we arrived—even if the anointing lifted. And so the song went on and on—and so did the road. But at last we saw a few lights ahead, and some familiar buildings upon a hillside—dear Bethany! We were very close now, for here was the Mount of Olives. The road circled around it and brought us into the city by the Valley of Jehoshaphat and over the Kidron Valley. All at once we were in the city—without having had that first stunning view I had hoped for. I was surprised when the bus stopped and we were told that this was our hotel. "Is this really Jerusalem?" I asked the porter as we entered. And one of our brothers said, "What else? We've been hours getting here." It was just about midnight, I noted. As one in a dream I followed Dwight and the porter down the marble hall, covered with oriental rugs, and into a small room that had a balcony. I put my bags down and rushed out onto the balcony. A large and very bright star was shining on a hill just opposite from us. And to my amazed delight I recognized the familiar outlines. "It is the Mount of Olives!" I cried. "And this is truly Jerusalem, the city of our dreams." As I said this another light flared on the Mount at the very top as though to signal, "Welcome!" Journey's end! From Wednesday morning until Friday at midnight it had taken us, and I was still wide awake, longing to keep an all-night watch on the balcony. Ancient words sang in my heart: "Out of the depths (Jordan's) have I cried unto Thee, O Lord." "My soul waiteth for the Lord more than they that watch for the morning: I say, more than they that watch for the morning."

THE CITY OF THE GREAT KING

y feet shall stand within thy gates, Jerusalem, Jerusalem. My feet shall
stand within thy gates, Jerusalem, Thou Holy City." How often had I
sung this little chorus in the past few years! And over and over again
my heart would leap with the Holy Spirit, as He had witnessed that it would surely
come true. And now I knew it was at hand—for somewhere nearby were the famous
gates of "The City of the Great King."

As I stood at midnight almost spellbound on our little balcony, I suddenly felt
King David at my side. And one of his Psalms began to sing within me:

> "I was glad when they said unto me, Let us go into the house
> of the Lord. Our feet shall stand within thy gates, O Jerusalem.
> Jerusalem is builded as a city that is compacted together. Whither
> the tribes go up, the tribes of the Lord, unto the testimony of
> Israel, to give thanks unto the name of the Lord. For there are
> set thrones of judgment, the thrones of the House of David."
> (Psalm 122:1-5)

I love the Knox version also:

> "Welcome sound, when I heard them saying, We will go in the
> Lord's house! Within thy gates, O Jerusalem, *our feet stand at last*;
> Jerusalem, built as a city should be built that is one in fellowship.
> There the tribes meet, the Lord's own tribes, to give praise, as
> Israel is ever bound, to the Lord's name."

David himself seemed to have come forth to greet me! I felt as though I was
looking out upon the city much as David must have walked upon his balcony and

looked out upon it thirty centuries ago. And I loved it as he did—as God does—passionately, tenderly and eternally. It lay before me now like a child, peacefully sleeping, after a busy day of work and play. "Thine eyes shall see Jerusalem a quiet habitation." This Scripture, and fragments of many others, flashed through my mind. After all, Jerusalem is mentioned approximately eight hundred and twenty times in the Bible! (Yes, Marian and I actually counted the references, and included its other names in the count.) It is likened to a wide variety of contrasting images: a virgin daughter and a menstrous woman; a burdensome stone and a cup of trembling; a lion and a den of dragons; a city of Truths and a heaps (desolation.) To the Psalmist it was "my chief joy," and the "joy of the whole earth." And to the Lord, "a praise in the earth." To an inspired modern writer, it is, "the sacred heart of the world." Yet to me on this holy night it was "my darling child"—a concept I have never heard used in connection with Jerusalem, of all places! I wanted to gather it in my arms and hold it to my breast. For the moment I had forgotten that this city had been overthrown by violent siege at least fifty times in its known history, and that it had been razed to the ground again and again.

> "Pray for the peace of Jerusalem; they shall prosper that love thee. Peace be within thy walls and prosperity (righteousness) within thy palaces. For my brethren and companion's sakes, I will now say, Peace be within thee. Because of the house of the Lord our God I will seek thy good." (Psalm 122:6-9)

On this night my "brethren and companions" were the precious saints of both the Old and New Testaments. And I felt a multitude of witnesses surrounding me as I lifted my eyes again to the Mount of Olives, thinking how often Jesus had looked thus upon it, had walked upon it, had prayed and agonized there, and would one day soon return in His resurrected body to stand upon it. I was almost entranced with wonder, for the star-like light continued to shine brightly. (I looked for it again on the two remaining nights we spent there, but it was not visible, so perhaps it WAS a special sign.) An old song we often sing at Passover time was quickened to me, somewhat altered in text:

> 'Tis midnight and on Olive's brow,
> The star once dimmed is strangely bright;
> 'Tis midnight in the garden now
> And all's at peace this holy night.

"Somewhere, up near that star, is the little new chapel which Brother Mattar is building," I thought. And I pointed this out to Dwight, who had joined me in gazing out upon the city. How I desired to prolong this priceless "enchanted interval" in which time seemed to stand still with me and look with rapture upon "the city of our solemnities," the Holy City of our God. Yet I was a tourist as well as a pilgrim, and must attend to the pressing matters at hand—the unpacking of our bags and preparation for a few hours' sleep. The morning would not tarry on its way. And we must rest so we could arise and be ready for all the day would offer. I reluctantly turned from the balcony and went back into our little room, but I have returned again and again to that midnight tryst in meditation. It is one of the most treasured intervals of our entire trip.

Almost everyone who makes a pilgrimage—or a tour—to the Holy Land has a very definite impression of his first view of The Holy City. I had read many accounts of how unusually thrilling it is to "go up to Jerusalem." And I had wondered just how I would go up—and what my own impressions would be. I knew, of course, that it could not be in the manner of the ancient Israelites who went up on foot, pausing to sing the Songs of Degrees and other Psalms at various places along the way. However, this is the way I would have loved it to be, for Jesus too, at the age of twelve, must first have gone up with just such a company. It is believed that He first saw the city from the top of Mt. Scopus—at the north end of the Mount of Olives. This is indeed a splendid place from which to view it. Nor could I go up on horseback, as many of the Crusaders had gone. It is said that when they reached a certain mountain from whence they caught their first view of Jerusalem, they were ecstatic with joy. They named it the "Mount of Joy" and it bears that name to this very day. Whether one came from the north, out of Samaria, or from the coastlands, or from the south, or from the interior, by the Jericho Road, one thing was certain—one always went *up* to Jerusalem. For, as is well known, the Holy City is set like a priceless jewel in a crown of golden hills and mountains which give it a special eminence and majesty. It seems a pity that today most of the travelers from afar arrive by air and catch their first view of the city from above, as they circle around to make a landing at the little airport which lies to the north, toward Ramallah. To *look down* upon Jerusalem seems all wrong; for centuries all men have looked up to it! Yet even this strange reversal may have a prophetic significance. The earthly Jerusalem is but a sign or type of the NEW Jerusalem which is from Above. And all men must look up and away from the earthly Jerusalem to catch a glimpse of it. However only truly dedicated pilgrims may see it from afar. Isaiah has told us: "Thine eyes shall see the King in His beauty: they shall behold the land that is very far off . . . Look upon Zion . . . Look upon Jerusalem!" This is one pilgrimage we all shall make. And it is certain that we shall go *up*, rather than down, to enter its gates. Praise God, it shall descend to meet us!

I have been truly thankful to the Lord for so arranging our tour that we too could go *up* into Jerusalem on that night of our arrival. And because it was so strenuous to travel all the way from Amman to the city, along with age-old Jericho Road (which the modern highway still follows) it was all the more rewarding to arrive at the mystic midnight hour and have a Davidic view from a balcony facing directly toward the Mount of Olives. "The hours I spend with Thee, my Lord (dear heart), are like a string of pearls to me. I count them over, every one apart—my rosary, my rosary!" This jewel is one of the most precious ones cherished in my memory-rosary.

As I have already said, I was a tourist, as well as a missionary-pilgrim. And this was one of the aspects of the trip I had most disliked from the moment I had first considered it. When one is on a Tour, all things are arranged and ordered in advance. And the schedule must be followed, no matter how one may feel about it. I have been so used to following the leadings of the Holy Spirit, letting Him graciously order my goings, that to be under any sort of set rule seemed distasteful. On the other hand, I realized that since we knew nothing about the people or the places we were to visit—except that which we had picked up by study—we would likely have found it very confusing to get clear guidance. When the Lord had made it certain that we were to go on the tour, He also made it plain that *I was to accept all that went with it as His permissive—if not perfect—will, and that He expected me to cooperate to the full, as gracefully as possible.* This required a constant discipline on my part.

The good tourist unpacks his or her bags promptly on arrival at the hotel, so that clothes can hang and wrinkles fall out. If the stay is for more than one night, one must carefully arrange clothes and toilet articles for convenient use. This I remembered, feeling suddenly utterly exhausted and longing just to flop on my bed. The little room had a marble floor and a small oriental rug. How kingly! It also had very hard mattresses with no real springs—just the kind found on cots. And it was small. There was not even a dresser, only a small nightstand. The one light was very dim, and it almost overwhelmed me to think of unpacking in such light without my glasses. But Dwight, meanwhile, was rejoicing in our large bathroom. We had been told that we would not likely have a bathroom in all the hotels. But oh how we needed one this night! Since all around us there was much commotion in getting our crowd settled, we felt free to run the shower and really refresh ourselves. By the time Dwight was showered and ready for bed, his overnight bag had been delivered to the door, and so too had my large suitcase. But his large bag, in which were his only good suit, my robe and other needed items, did not arrive. So his unpacking was brief, and he was soon sleeping soundly in his bed.

About this time I began to get into a sort of nightmarish state. In my exhaustion and eye-strain, I seemed not able to locate many things I needed—among them my curlers and bobby pins. The care of one's hair is a problem on a trip, so I had arranged to have mine permanented in such a way as to require minimum time and effort to shampoo and set. The success of my coiffure depended on my larger roller curlers, bobby pins and small combs to use in the back. All of these I had carefully placed in a plastic bag and packed. At least I thought I had! But they were nowhere to be found. And my hair by this time was really a sight—for wind, rain and my hat had all contributed to its derangement. (Yes, this is exactly the right word!) I simply had to do something about it that very night. The contrast between the divine contemplation of Jerusalem and the deranged manner in which I began searching for my curlers is really laughable now, and typical of our "mortal miseries." But that night I was nearer tears than laughter after a fruitless turning-upside-down of our bags. By this time other things too were misplaced and I seemed helpless to bring order out of confusion. Dwight awoke and urged me to get to bed. But I was still waiting for the porter to bring his large bag to the door. Surely in it I would find my hair things! But at length, still frustrated, I had to give up—for the bag never did arrive, and I was not able to keep awake longer.

All over my body I had found irritated red places where my clothes had rubbed me constantly. And some plastic-coated garters had apparently poisoned my legs, for they were sore and angry-red where the garters had been. I crawled into bed feeling like Mrs. Job, oh so miserable—hair stringing, body covered with sores and aching all over! Sister body was taking it hard! But Sister spirit was still singing and thanking the Lord for His goodness in delivering her safely within sight of the city gates! My last prayer and thought was, "Please let me wake up early, so I can go out and meet the dawn with You, dear Lord." It was about 2.30 when I lost consciousness.

Three hours later I was gently awakened by an unfamiliar sound which I was to hear often in Jordan. I can close my eyes now and hear it clearly, though I don't know how to put it on paper. It was always a man's voice that I heard—or a combination of men's voices—and the little strain that is half chanted, half sung, would vary, yet be strangely similar. Perhaps this first singing call I heard was from a Moslem at prayer. "A-hay-a-ah-a-hay"—in a minor key. It covered a range of tones from high to low and was repeated over and over. Later on I was to hear it from the peddlers pushing carts in the street, from the workmen carrying cement up the ramp of a building, or from a boy walking in a school yard and reciting from a book. I believe the Jordan Arab sings what he wishes to say, and this goes on all throughout the day. Though it is in a minor key, it sounds very happy indeed. I responded to their song—and loved it from the first. I even imagined, on that blessed morning, that Jesus was calling me in Arab fashion to arise and meet the dawn with Him. Three hours' sleep does not make up for three full nights—so I was very tired in body. But I was truly eager to be up and out. And I recalled a saying credited to the early Christians, who were said to have greeted each other on Easter with the stirring phrase, "He's up and out!" I was too—at least I was up and out on the balcony. The sunrise was at hand, for the whole city was well lighted. Just next to the hotel was a school, and several boys in their teens were walking up and down reading from their books, paying no attention to one another. Men were stirring about in the streets below. And the Mount of Olives was even nearer and clearer by day than by night. It just couldn't be real—but it was! I hoped I could dress quietly, without disturbing Dwight. So I tiptoed about and managed to be fully dressed before he stirred. I told him I was going out just for a short walk, and since he knows I want to be alone with the Lord in the morning, he made no protest.

I picked up a long white scarf to wrap about my tousled hair and shoulders. It was so warm already that this was all I needed. Very quietly I slipped down the hall, grateful for the rug to muffle my steps. I was just about to push the elevator button when a rather surprised porter appeared, smiling and bowing, and saying, "Please, Madamé?" (This phrase is used constantly in the Middle East, and means any one of a variety of things, depending on the circumstances. It became a by-word with all of us. Madamé is always said with the accent on the "dahm.") I knew he meant what did I want. Well, obviously, I wanted the elevator. But I did not want to be rude, so I decided to attempt an Arabic greeting. This surprised him even more. He knew what I meant, but repeated it to help me get the accent just right. This started up a rather ridiculous early morning conversation, partly in English, partly in Arabic. It took a while to make it plain to him that I wanted to go out and see the city. He looked very distressed, then suddenly burst into smile and said, "Yes, Madamé, I show you." This was just what I didn't want—I was tempted to be impatient—for I was eager to get out and be alone with the Lord. Then, to my surprise, he took my arm most tenderly and led me away from the elevator toward the end of the corridor. I thought it must be that the elevator was not working, so I must use the stairs. Then he opened a door and began to take me through a disorderly little room, which I suppose was used for storage. This really shook me a little. For one wild moment I thought of weird stories I had read about Arab treachery and intrigue. Just where was I being taken? I wished strongly for Dwight. But I did not want to show my fear or suspicion, remembering that I was on a missionary journey—and here was a precious soul. We crossed the room, going around a few chairs and bundles, and went up some stairs to another door. Most mysterious! Then, after making a turn, we were out on a large flat roof. And of course I understood—this was his way to show me the city! (Was it Jesus in Arabic form?) Still holding my arm, he escorted me about, pointing out the Mt. of Olives, the walls of the city and the Dome of the Rock mosque, now a bright gold in the rising sun. I had no idea what it was, for I had seen it pictured only with a lead dome—all the gold having been removed and used a long time ago. "Is that really the Dome of the Rock?" I asked him, and he assured me that it was. (Later we learned that only recently the dome was gilded again, in fact just before our arrival!) Someway it was much smaller than I had expected. I was startled to find it so near—only about two blocks away! After touring the roof, I tried to let him see that I still wanted to go downstairs and out into the streets. He again seemed so pained and confused, that I began to wonder if perhaps a woman was not supposed to go out alone so early. This must be it! So I said, "*Shookran* (thank you), and that was *Qwa-ese* (fine, good.) Now I must go and call my husband, for we are going out." He seemed to ponder this, so I pointed to my wedding ring and then pointed out, and soon he laughed and agreed. Again he escorted me all the way down the stairs and through the room and into the hall. Apparently he was only being friendly, but I was relieved when I was finally on my own again.

Of course I got to Dwight as fast as I could. I found him almost dressed and eager to go with me. I waited on the balcony, while he finished getting ready, and I was amazed to see that already the streets were filling with people, even though it was only shortly after six. At the elevator the same porter met us, and we went all through the Arabic routine again, he and I. Every delay seemed most trying, but finally the elevator came and we got to the main lobby. The man at the desk also appeared startled to see us so early. He too tried to delay us. I began to wonder if they had some law about tourists not getting out to see the dawn in Jerusalem. We were as excited as children when we finally hastened down the street. I wanted to find that golden dome, and we had to circle around a very winding, narrow street, on which were little shops, until we came upon an open place where two or three roads converged. It was actually swarming with people, sheep and vehicles at this early hour. I saw no women at all—only men and boys. And I was very thankful indeed to have my husband's strong arm to lean on, for they all stared at us, then smiled and spoke when we did. I could be quite free in beaming on them, and in calling out "*Sabah Al Khary!*" (Good morning!) When they heard this, they usually laughed and responded with seeming pleasure. Later I learned that not many Americans made an effort to learn their language, and if one does, they are truly pleased.

We noticed a large gate just ahead of us, and beyond it, we knew, lay the Old City. We felt to turn left and follow the old wall, rather than enter it. But we wondered which gate it was. All along the old wall were men and sheep. And rightly so! For this turned out to be Herod's gate, and it is now used as "The Sheep Gate." We came to a corner of the wall, and followed the road as it led on downhill. Several men were working with beautiful white limestone, cutting it for a building. I remembered that it is taken from the quarries under the city, and that it is soft until exposed to the atmosphere, then it hardens and becomes very strong. As I looked beyond the men, I did not realize that I was gazing across the Kidron Ravine, where once the brook had run, and that, just to our right it merged in the Valley of Jehoshaphat. As carefully as I had read, studied pictures and tried to learn all about Jerusalem, I felt entirely ignorant and confused—yet very much at home. We soon found a boy who could speak a little English, and we asked him about the beautiful church just across the way. He seemed astounded that we didn't know that this was Gethsemane and its Church of the Nations. Imagine my surprise! And I had seen the picture of it many times. Yet it seemed to me that my uncertainty was partially because of my inability to grasp the fact that at long last we truly were walking in the real Jerusalem, not merely looking at pictures or seeing it by the Spirit from afar. When we came to a turn in the road, we felt to go up, instead of down, and I saw something then that I did recognize—the old "Golden Gate"—the "Gate Beautiful" through which Jesus made His triumphal entry into the city on Palm Sunday! This is the gate the Moslems closed, since they have a prophecy about the Messiah entering the city again through it on some future day and ending the Moslem dominion. The road passed directly under this gate, and I stood there awhile looking up to it, praising and adoring the Lord. In front of it is a Moslem cemetery and there were weeds and stones lying about. But in the midst of this debris were some beautiful little scarlet anemones gaily dancing in the morning breeze. This flower, the real "lily of the valley," has been especially dear to me since early last spring, when the Lord spoke to me about it in various ways. I had so hoped to see a few still in bloom—though I knew that they appeared early in the springtime. And here they were, the first I saw, just across from the Garden of Gethsemane, beneath the Golden Gate!

Not until after we returned home did I realize that we were walking over the very part of Kidron Ravine into which the stones of the Temple had been toppled, after it was destroyed. Beneath our feet lay a great mass of ruins, not only the sacred Temple, but the debris from other buildings lay there too. No wonder I felt so awestruck and thrilled!

Dwight, meanwhile, was trying to get a few pictures. About this time I saw a group of schoolgirls coming toward us. It is still hard for me to figure out why school children were out so early! They were shy, but one who knew a little English lingered to talk to us. I saw that Dwight was as thrilled as I. Everything we saw delighted us. We felt like little children again, and kept exclaiming loudly, voicing our joy. It seemed to me that we were both given a real "baptism" into Jerusalem that morning. It was a delight to me that Dwight was in complete harmony with me in the Spirit, for I knew he had not made the study and preparation we in the Company had made. Tears and laughter mingled as we turned around and slowly made our way back toward the hotel. We were indeed like "those that dream."

A handsome English-speaking Arab boy attached himself to us, and courteously accompanied us for several blocks. And a kindly old stonecutter hastened toward me with a handkerchief I had dropped as I had passed him earlier. How I loved those Jordanians! And still do! Majella's dear song about Jesus waking in Jordan Land was sweetly brought to mind as we turned away from the old walls of the city. Sometime, during this very day, I felt sure that we would enter them and stand with Him within the city gates.

WITHIN THE GATES

We had gone out of the hotel into the dawn with quick steps and high hearts. When we returned our hearts were still high but our steps were dragging. Even though I had worn my good "health shoes" my feet were hurting! I had read about those uneven pavements, cobblestones and slippery places, and it is all *so* true! One needs heavy rubber ripple-soled sandals. Fortunately I had some along. When we entered the hotel we seemed to leave behind the "wonder-world" of dear childlike Jerusalem, and to enter a jumbled world of upset adults. Several suitcases had been lost in addition to Dwight's. And there was a lot of loud talking around the desk and in the lobby. I pitied the women who had lost their bags containing their walking shoes and cool dresses. It was very hot by now, and I suddenly felt exhausted, old and a bit fearful. I didn't quite know why. So I hurried to our room, while Dwight waited to see about his bag. I planned just to collapse on the bed for at least ten minutes. But when I finally got the door unlocked—it protested and stuck for a while—I came near collapsing on the floor than on the bed! I held onto the door and stood aghast for at least a minute. In our absence our beds had been made up, and all the vast disarray of our bags—through which I had searched wildly for the curlers—had been picked up and stacked neatly on one bed. And my purse was nowhere in sight!

I was overcome with dismay. First, I think I felt humiliation. I had not expected anyone to make up our beds at seven in the morning! Instead, I had supposed that sometime later, while we were at breakfast or out sightseeing, this would be done, as is the custom in America. I would never have left such an untidy room for a maid, had I known. My second emotion was stronger—fear and self-condemnation for having left my purse with my money, passports (which they tell you NEVER to let out of your sight), and travelers checks in the room. In addition to this I had left a tiny "bosom-bag" in which I had placed some additional bills. I had worn it while traveling and had removed it and completely forgotten it in my excitement

107

and exhaustion. I had also left one or two pieces of rather valuable jewelry there. So you see I had broken another vital rule for travelers: "Never leave money or valuables in your room, even in a locked bag—either carry them with you or place them in the hotel safe." How could I have been so stupid and forgetful! All the years of training about "practicing presence of mind" should have resulted in making me a thoughtful, careful traveler. I reproached myself for "going off the deep end" to such an extent that I had little sense. Locating my purse, I fearfully opened it. It was a relief to find the wallet, money and papers all intact. But where were my travelers checks? Had I moved them or what? I turned to my suitcase, and soon had it topsy-turvy. There was no trace of the little bag or checks. I tore through the overnight bag—but could not find them. Suddenly I wanted to cry my eyes out—I even felt a little sick. Just a few minutes before I had been overflowing with love for these dear Arabs. They had seemed so childlike, honest, kind and friendly. But apparently one of them had robbed me! I think we all know the shock of suddenly discovering a betrayal of love or confidence. I felt heartbroken about it.

At that moment Dwight came in and announced that they had found no trace of his bag. All the training and experience through the years in suffering loss and taking joyfully "the spoiling of our goods," stood me in good stead in that awful moment. My glasses were gone; my money, checks and some jewelry were stolen; Dwight's bag and his best suit and clothes and some of my best things were missing. It looked as though we were having strenuous exercises and being graceful losers. I determined to joy in God, nevertheless, and decided not to say anything to Dwight about my losses until later. He was eager to eat breakfast. So I hurried to clean up to go with him. After all, I told myself, I had no one to blame but myself. All these Arabs were hopelessly poor, and no doubt they needed what had been taken. I could report the lost checks and would eventually have a refund. The bag and contents were insured. But, for the rest of the trip, I would surely be short of money, though Dwight too had some. And how we would miss our good clothes! As we got to the elevator I saw the friendly porter again. And I wondered about him. He had known that we were both out, and he had a key. There was a maid there too. I forced myself to be calm, and asked her if she was the one who had made up our room. The porter interrupted to tell her I could talk Arabic—and we went around in some more circles.

But finally, in French, I got her to say that she had done so. Then I thanked her and told her that I wanted to apologize for having left it so disorderly. She laughed and said, "Everybody do, not only you; no matter!" Was it she who had taken the things? What should I do—go to the management, send for the police, or just trust the Lord? Instantly I decided on the latter and off we went downstairs.

Breakfast was a nice surprise! And about that time I needed one. I was shaking and very weak, and felt as though I hadn't slept for nearly a week, which was just about the case. And I was hungry! Before long a better breakfast than the continental one, which we expected, was set before us, nicely served in an attractive dining room. It was mostly eggs, bread, jam and coffee. But it was very good and there was plenty of it, and the waiters were charming. I began to regain my senses and to get back into conscious communion with the Lord—my thoughts seemed to have been staggering about in my mind. How vividly the Lord reminded me then of dear Dr. McKoy's loss of both wallet and suitcase, when he first went to India. I could appreciate all the more the grace and faith it took for him to rise up and go on. Experience surely does stimulate empathy!

By the time we returned to our room, I was reconciled to just take the loss quietly and do nothing outward at all. I realized that had I consciously prayed about my money and asked the Lord to look after it, I would have remembered to take it with me. I was *not* fully depending on Him in *everything*, it was plain to see. We had only about fifteen minutes in which to get ready to go on tour. But I managed to go through my bags again. And to my amazement, one by one, I found every missing thing! To this day I cannot understand it. And I still do not know if they were taken and later returned, or if they were never taken in the first place. (Several times in my life I have had things completely disappear under mysterious circumstances, and then reappear in the same manner. I feel the enemy either "hides" things from us, perhaps by oppressing us in some way so that we cannot see them; or that there is some truth to the idea that evil forces can cause the loss of material as well as spiritual things.) In any case, I truly praised the Lord, and thanked Him over and over. And I resolved to give those attendants a good tip when I left. How awful it would have been had I rushed to the management and made trouble! Thank God for training in *patience* and *waiting* before *acting*!

We started off on our tour with eager hearts. But some of our crowd were still fussing about lost articles. A few of the women were fuming because we had to walk to the city gates, instead of being taken in a bus—though it was only about two long blocks! (Vehicles cannot operate within the walled city—except on one street which they may now traverse. However it is too narrow for buses to use.) Our guide was a well-educated and rather clever Arab named Ahman, and I am sure he was thoroughly acquainted with Jerusalem and environs. But it was difficult to understand his broken English, and one had to step lively to keep up with him. Only a moment was given in which to stop and look and try to grasp what we were seeing and what he was telling about it. Taking notes was out of the question, for people thronged about us: our own crowd, other tourists and the children. Always dear, dirty, smiling children, with their eager "Halos!" Thus being jostled about, trying to hold on to my purse, returning their greetings, giving out little Jordanian coins—which Dwight had obtained—to little outstretched hands, in addition to trying to follow and listen to the guide was really almost too much for me! But I did the best I could. I had to run to catch up with the guide as he turned in toward Herod's Gate. I realized now that our hotel was located to the north of the walled city, and that we were making our first entrance into the city from one of the north gates, called Herod's Gate, and now used for *The Sheep Gate*!

Oh how very dear to my heart are the ancient gates of the Old Jerusalem! I well recall how the Spirit led us to make a study and write a most beautiful message about these twelve significant gates, which were rebuilt into the walls in the days of Nehemiah. The Sheep Gate and the Fountain Gate were especially significant to us, and our Lord instructed us that our going up to the Heavenly Jerusalem must be through THE FOUNTAIN GATE. "And at the Fountain Gate . . . they went up by the stairs of the city of David . . . above the house of David eastward . . . and the singers sang loud . . . . Also on that day they offered great sacrifices and rejoiced; for God had made them rejoice with great joy . . . so that the Joy of Jerusalem was heard even afar off." (Neh. 12:37-43) How truly wonderful and real was our prophetic going up into the city through the Fountain Gate—a type of our greater going up into the New Jerusalem by the "Fountain" way. How can I ever forget His glorious dealings at that time, and even earlier, when He led two of us to "go out by night" and compass the

Coliseum in Los Angeles, likening it and its gates to the walls and gates of Jerusalem. Truly He walked with us that night! And what fervent prayers arose! But all this was in the past. So I tried to stir myself to realize that on this holy Sabbath (Saturday), May 13, 1961, I was about to enter the Holy City in the flesh on behalf of our Lord and all of you, as a witness and worshiper.

It was a most undramatic entry! No banners were raised! No trumpets blew! No Sons of Israel stood at attention—(Indeed none are permitted even to set foot in Jordan.) And no one seemed to realize what a tremendous moment this was! Picture me, if you can, almost shoving my way—through the jam of tourists and natives—on into the gate. About that time I heard a sheep go, "baa!" And I felt like one indeed! (I also felt that I was a Jew, representing all Israel, to whom the city is now closed.) Our guide actually "herded" us on and we picked our way through the rough, narrow, crowded street. I got as close to him as I could and heard him say that St. Stephen's gate was just ahead. He told us that the Arabs call it "Mary's Gate," "*Bab Sitti Maryam*," since she was born nearby. It has also been called, "The Tribe's Gate," and "The Lion Gate." Near this gate, outside the walls, St. Stephen is said to have been stoned; but others believe that the stoning took place just outside the Damascus Gate, near Golgotha and the Garden Tomb, where Jesus died and was buried. Our thoughts were turned from Stephen to Mary, as our guide led us through a narrow doorway and toward the Crusader Church of St. Anne, the supposed birthplace of Mary. The home of St. Anne and St. Joachim, her parents, was partly hewn out of rock, and this forms the crypt of the church. A Christian church was first built here in the 6th century, and it was later rebuilt by the Crusaders. A Biblical Museum is located on the right. We were led down into the crypt where the infancy of Mary is depicted in little statues. Many of our company seemed confused and thought this was supposed to be the manger scene. They began to fuss about being taken to Catholic Churches and shrines. I was surprised that so few, even among the women, seemed at all interested in the birth of the Virgin Mary, the precious Mother of our Lord, and our own "Mother-in-Grace"—as He once said to us. I dared to kneel and honor her there, and the Spirit sweetly rejoiced in me. And I left in a niche of the crypt our first little souvenir token, which, as it happened, had been given to me by Majella. I was the last to leave, and had to run to catch up with the party now moving on toward the newly excavated Pool of Bethesda. For centuries the location of this pool, mentioned in John 5:2-9, was lost. (As you no doubt realize, all of Old Jerusalem lies buried under the present city—from 20 to 60 feet.) In 1871 it was rediscovered. Excavations beginning in 1956 have uncovered not only the pool but also parts of Byzantine and Crusader churches. With thankfulness and awe I descended the many steps down to the pool. Here was a place where Jesus had *really* walked and worked a miracle! The last part of the descent was very slippery, so I waited above while Dwight went down, dipped his hands in the water, and returned to lay them on my head. Even though I am a "good" Baptist, I felt that I was sweetly rebaptized that Sabbath morning in the clear cool waters of the healing pool of Bethesda. With joy I noted the Bible story written there at the entry in some 40 languages. I could not restrain my tears and praise as I walked about, looking at the unfamiliar languages in which such familiar words were recorded. "These are the languages of our dear nations," I thought, and longed to speak *all* of them fluently. (Of course the Holy Spirit *can*, even if we can't!) Near St. Anne's was another famous pool, called Mary's Pool. There seemed to be some sort of bathhouse there, and one could bathe for a fee. How I would like to have done so!

By this time most of our party had finished taking pictures and exploring these amazing ruins. In some ways they were the most satisfying of anything we saw within the walls, for we knew that almost every other thing we saw was erected there after the time of Christ. The one exception is the Wailing Wall, a portion of which actually dates back to the Temple of Solomon. Our guide now directed our steps toward this in the Temple Area, telling us many interesting things as we walked. The present walls of Jerusalem, we learned, were built in A.D. 1540-1542 by Suleiman the Magnificent, during the rule of the Turks. This Sultan was highly spoken of by the guide. Most of our group thought he was talking about Solomon, and seemed not to notice the discrepancy in the dates. The walls average 38 feet in height and enclose an irregular quadrangle about 2½ miles in circumference. The Temple area occupies about one-sixth of the walled city.

The old Temple site, now called *Haram Al Sharif* (The Enclosure of the Chief Sanctuary) is truly impressive. It was nearing midday as we entered it. After walking about in the closely compressed city streets, it seemed all the more amazing to come upon this vast, uncluttered court-like area which Dr. Field describes as a "great, level, stone-paved park, broken by the central raised platform." On the lower level there is a cistern, one of the many underground reservoirs from which water is drawn up into the city. Among them is one called "The King's Cistern," 40' deep and over 200' in diameter—said to have been there in Jesus' day. There was also a pool, a place of ablution for the Moslems, who must wash their hands, face and feet five times each day before bowing toward Mecca in prayer. Two or three were doing so as we drew near. But there were few people in this area, and everyone was quiet and reverent. Even upon our own crowd peace seemed to settle down. By this time almost everyone was very tired and also very hot. I longed to splash around in that pool and to get a drink—especially to get a drink!

As we slowly ascended about twelve feet toward the Dome of the Rock (erroneously called the Mosque of Omar), most of us were silent. Perhaps we were trying to picture in its place the beautiful Temple of the Lord which once stood on this hallowed ground. Our guide, meanwhile with obvious pride, was reeling off facts and figures about this very famous building. Some have said that it is second only to the Taj Mahal in beauty. Personally, I found it very majestic and inspiring. But to describe it adequately would indeed be difficult. It is octagonal in shape, and its sides are covered at the base with marble and on the upper part with fine blue and white porcelain tiles, on which are inscribed arabesques and verses from the Koran. A magnificent dome, now brightly gilded, crowns the mosque. Our guide explained that this is the religious center of the Muslims of the Middle East, second only to Mecca. They believe that Mohammed made a miraculous journey to Jerusalem, and that he ascended into heaven from the famous Dome of the Rock over which it is built. The mosque is the earliest example of Arabic architecture in existence today, having been completed in A.D. 691. Under the Crusaders it was converted into a church for some 80 years. Saladin restored it to Moslem worship in 1194. It has frequently needed repairs, and at present an extensive program of restoration is being carried on. (These many facts I could not possibly remember accurately, so I purchased a little book to remind me of them.)

I felt a special thrill as I slipped off my shoes and followed the guide into the interior. (Large flat canvass over-shoes are provided for visitors, for none may enter in their street shoes. However there was a lineup—others of our Tour, who had another guide, had arrived ahead; so our guide suggested that we not wait, but just go in stocking-footed.) The four doors are at the compass points. The north door is *Bab al Jannah* (sometimes called the Door of Paradise), and the eastern one is called *Bab Daoud*, Door of David. I believe this is the one by which we entered. The interior of the Mosque was fabulously beautiful in an Arabian Nights manner. Costly Persian rugs, fine marble, exquisite tiles with intricate arabesques, and rich gold blended in regal oriental splendor. I was almost overwhelmed and I responded to this beauty in a special way, I believe, because for so long the Spirit has dealt with me about the Moslem peoples, and given me a rapport with them. It is said that no matter at which of the four doors you enter, the building is so ingeniously constructed that you may see at once *all* the pillars and columns without moving about. There are eight pillars with two marble columns between each pair, forming the first octagonal concentric enclosure. The inner one is circular and is formed of four pillars, with three taken from the Byzantine buildings destroyed by the Persians in 614. The inner decoration of the dome is of elaborate mosaic. There are no statues, altars, pictures or clutter of religious objects. The Islamic faith permits no representation of God in any but symbolic form. Stars, rosettes, lotus, lily, trefoil, tulip, and the *cross* are used. The effect is one of majestic spaciousness and peace. Some have said that a mosque is cold, in comparison with the warmth of the colorful Roman Catholic churches. But I found it almost celestial-like in atmosphere. Filtering through the stained glass and opaque windows, the sunlight itself seemed more heavenly.

The heart of the mosque is the Sacred Rock, emerging from its foundation—massive, gaunt and polished like a huge piece of coal. It is irregular in shape, about 58' long and 44' high. This is said to be the natural altar on the summit of Mount Moriah on which Abraham purposed to offer Isaac. David bought this land from the Jebusite, Ornan, and raised an altar here. In the time of Christ it is thought that the altar of burnt offerings was built just over it or near it. Because so many pilgrims took pieces of it away, the Crusaders erected an iron grating around the rock to protect it, and it is still there. Under the rock there are said to be several praying places. As we approached this rock, I walked behind the others. I have always felt a special interest in Mount Moriah and this rock, and I wanted to approach it alone and in the Spirit. I was a little shaken when I saw a distinguished Arab in native dress move swiftly toward me. A strange feeling came over me—a sense of "déjà vu"—of having previously lived through this moment, or at least having seen and known this man. He gently took my arm, and in perfect English softly spoke to me of the significance of the rock and mountain, on which the Temple—and now the mosque— was erected. I felt something special in this mysterious stranger—it was as though the Lord Himself drew near, took my arm, and spoke to me out of His heart. And now it is even more real to me as I write, than it was that day! How I praise Him! Who was this man in native dress? Apart from Revelation, I shall never know, for as suddenly as he drew near, he left me, murmuring graciously, *"It faddal"* ("be gracious," used for "excuse me" etc.) As in a daze I walked around the rock. Then I hastened to catch up with our group who were being led by now closer to the *mithrab*—a niche or recess in a mosque denoting the direction of Mecca—and the *mimbar*, the pulpit from which the Koran is read. While we sat on the luxurious rugs—the most costly, I believe, I have ever seen—our guide explained to us the basic tenets—five pillars—of Islam. Most of our group were courteous, but a few were annoyed and plainly showed it. It was apparent from the comments and questions that few knew even the basic facts about Islam—one of the world's leading religions. After we arose, grateful for a short rest, Ahman told us that anyone who walked around the *mimbar*—a most intricately carved and beautiful one—would receive from Allah whatever he wished for. I felt at once to make the little circle and wish a wondrous wish—that all the Moslems in the world would receive Jesus Christ as their Savior, as well as honor Him as a prophet, which they already do. I recalled that written on the walls of this mosque is a quotation from the Koran which reads: "Jesus says, blessings be on me on the day of my birth, and of my death, and of my resurrection to life . . . . He is Jesus, the Son of Mary, the Word of Truth, concerning whom some are in doubt . . . . When God hath resolved upon anything, He says, Let it be, and it is." But there is another quotation that says in contradictory fashion: "The Messiah Jesus is only the son of Mary, the ambassador of God, and His Word which God deposited in Mary . . . God is One, and far be it from Him that He should have had a son." Well, if my wish comes true, they will *know* God had a Son, and that He Is Jesus!

I wanted to linger longer in the mosque, but after secreting another little token, I followed the guide out, found my shoes and went on with the group toward the Wailing Wall. Enroute we briefly visited the very large *Al-Aqsa* mosque, which is almost as sacred as the Dome of the Rock. This mosque has a special chapel for women. The entire structure is magnificent with marble and rose-colored limestone, stained glass windows, many pillars and a most elaborate ceiling. The small Mosque of Omar lies at the southeast corner of *Al-Aqsa* and is connected with it by an archway. There is also a small mosque which is a sort of model of the Dome of the Rock. Perhaps there were others too, but we did not visit them. We were eager to get to the Wailing Wall. Although our guides were most gracious toward us and Christianity, they showed some spleen when they spoke of this famous wall. This was the place where the Jews mourned for centuries, and read aloud the Scriptures about the restoration of Israel. It was evident that the Arabs were delighted that no Jew may now get anywhere near it. I touched it, but did not feel to wail. We marveled at the huge stones in it and thought again of Solomon's temple, of which it was a part.

By now it was lunchtime and all of us were ready to return to the hotel. Our guides thoughtfully provided us all with a cold drink, and the drink stand was conveniently located by a gift shop, into which, to my surprise, most of our crowd disappeared. (Later we learned that the guides owned an interest in it.) I was a little shocked that so many were thinking of souvenirs and novelties so soon after visiting such Holy places. The sounds of bartering carried out into the street, where we sat down to rest. Friendly Jordanians and visitors from other countries surrounded us, and we truly felt at home and among friends. Street peddlers quickly arrived, and soon there was a whirl of confusion. All the impressions of the morning seemed to fade out before they had time to "set." How I longed for a quiet place to meditate and make a few notes! But none was available. When the shopping spree finally ended we were all led to the Damascus gate and on out to the waiting bus. Oh welcome sight! Anymore walking seemed almost impossible at that moment.

As we rode toward the hotel I tried to recall what the various gates were. But later I had to look them up. There are 34 towers and 8 gates. The New Gate, the Jaffa Gate and David's or Zion's Gate, are presently closed, as they border No Man's Land.

The Damascus Gate (*Bab el Amud*) is the largest and most impressive of the gates in use. Then there is also Herod's Gate (*Bab Ez Zahira*), St. Stephen's Gate, and the Dung or Moor Gate (*Bab el Maghariba*.) The Golden Gate (*Bab Ed Dahriyya*), as I mentioned, is completely walled up. It has been said that it is possible to walk almost entirely around the city on the walls. How I would love to have done so, for we have often been stirred by the 48th Psalm which says: "Walk about Zion, Go round about Her. Tell the towers thereof. Mark ye well her bulwarks, consider her palaces." I could hardly wait for our next excursion within the walls. There was so much we had not even glimpsed as yet!

THE RIVER OF DEATH
AND THE DEAD SEA

T o be baptized in the Jordan River is the dream of many a devout follower of Jesus. Every year there are many pilgrims who put on their baptismal robes, stand upon its "stormy banks" and descend into its waters. However, the Lord did not put this desire in my own heart; though had I been traveling with a small company, such as our own, I might have felt such an urge. In a certain sense, I guess I felt that I had already experienced such a baptism—for a number of years ago, at the beginning of my Rapture Experience, the Jordan overflowed its banks and ran in our midst here in California. The Holy Spirit made very real to us at that time the symbolic meaning of this river. We had vivid dealings about the crossing of the Children of Israel. We saw how the priests were the *first to enter* the swollen waters, and the *last to leave*. Bearing the precious Ark of the Covenant, with praise and faith, they stepped into those treacherous waters, which even the strongest swimmers find difficult to ford, due to the undertow. We saw how the Lord turned back the flood and held the waters with firm control "while the priests of the Lord . . . stood firm on dry ground in the midst of Jordan, and all the Israelites passed over on dry ground, until all the people were passed clean over Jordan." The same Israelites who had passed over the Red Sea—symbolic of sin—and were baptized figuratively into it, were now baptized in Jordan—the symbol of death—and were passed over into a new life in God. We were shown how the Lord also told Joshua to choose a man for each of the twelve tribes of Israel. These men each took one stone from the dry river bed and carried it to the other side, so that a memorial of the miraculous passage might be erected. As the Spirit reenacted these things in our midst, we felt His awesome presence in an unusual way. I shall never forget how He called me into the waters, as a priest of the Lord, and baptized me therein, in the name of our Joshua, the Lord Jesus Christ.

Meditating about this now . . . and then turning my thoughts back to that holy Sabbath in Jerusalem, I can see how fitting it was that our first day in the Holy Land was "Baptism Day." Biblically speaking, our testimony as a Christian begins when we openly follow our Lord in water baptism. This is the first outward step from which all else proceeds. This is our public confession that we have entered into the death, burial and resurrection of Christ. (Although our actual salvation begins when we repent and believe on our Savior.) A similar pattern was followed by the Spirit on this blessed day in Jerusalem: early in the morning I was baptized in the sweet childlike "spirit" of the city; in midmorning I was baptized with the deep, clear hidden waters of the healing pool of Bethesda; and before the day ended, I was to be taken to the waters of the most fabulous and famous sea in the world, the Dead Sea. As we were returning to the hotel after our morning tour, our guide urged us to eat our lunch at once and get ready promptly for the long afternoon pilgrimage to the Jordan and the Dead Sea.

The announcement stirred up mingled feelings within me. I was longing for a short time to rest and recollect the myriad impressions of the morning. When I had first been certain that I was to go to the Holy Land, I had considered taking with me a small, portable tape recorder to capture the sounds and voices of the people, to record things the guides would tell us, and to talk into it myself, relating impressions as we journeyed. Yet the Lord had checked me from doing this. It seemed that He wanted me to be free to be in the Spirit and to do or say what He desired, rather than become involved with some mechanical device. Dwight, of course, would be the cameraman, gathering pictures we would all enjoy and cherish. But all my pictures and impressions were to be recorded on that amazing "built-in" camera and recorder with which the Lord has endowed us—memory. How I praise God that He began training my memory in very early childhood, even at my mother's knee. And throughout the years He has continued to stir me to memorize many things. It is a power that increases as it is used! In our midst we have had repeated dealings about consecrating and using memory for the glory of God. All this training in "recording," recollecting and re-expressing the various movings and messages of the Holy Spirit has greatly enhanced my ability along this line. And truly, dear ones, were this not so, such an account as this could never have been written. Already, though I have been home only five months, the events of the trip are fast fading into the subconscious realm of my mind, and I feel an urgency to hasten and transpose them into writing, before they are lost.

Several factors play an important part in our ability to remember. First is the power of *keen observation. We can never recall what we fail to record.* Personally speaking, I had to train myself a lot in this facility—for I am inclined to become detached and blind to many things. *Sensitivity to impressions* is needful too, for if we are dull and unresponsive, no deep impression can be made. Many of us go through life half-asleep, just taken up with our own little egocentric thoughts and activities. "Who is blind, but my servant? or deaf, as my messenger that I sent?" So speaks the Lord, and how deeply did He convict me of my blindness and unresponsive nature! Only after much prayer that He would open my eyes and my ears, and sensitize my whole being to feel, respond and understand, did I become alert and sensitive to the things of the Spirit. "Hear, ye deaf; and look, ye blind, that he may see!" I recall now how hurt I was when an anointed brother once spoke these words to me and prayed that God would open my eyes and my ears. But he was oh, so right, as I later learned! Yet I had supposed myself to be very spiritual, as I had received the baptism and had wonderful experiences. For that matter, I am still praying for *seeing eyes, hearing ears* and an *understanding heart!* Just learning *how to listen is an art.* Diplomats and secret agents are given extensive training in *listening,* in observation, and in *remembering correctly* a multitude of things which they do not dare to put down on paper. Yet we, the secret Agents of the Kingdom of God, sometimes fail to realize the importance of such training and discipline.

During this trip the Holy Spirit often impressed me with *the great value of a good memory,* utilized by the Holy Spirit for the things pertaining to the Kingdom of God. I cannot too greatly stress it! Another factor is the making of clear, concise notes. It is said that one clear memorandum is worth an hour of recollection. But *recollection too plays its valuable role.* I have learned to get quiet as soon after a meeting or event as possible. I attempt to picture again all that took place, to rehearse what was spoken or revealed, and to make *"mental notes"* about it. The important factors in memory then may be summarized as: keen observation, discernment, or sensitivity to impressions, ability to listen and record what is heard, recollection and making written memos. *Yet all of these factors were impeded on my trip,* for there was so much to observe that the pictures seemed to blur and run together. My impressions were so multitudinous that they were shallow and fleeting. Even though I listened intently, I could not recall anything clearly—for so many sounds were around me. And there was seldom a moment to pause and take a "time exposure" of any thought, let alone quietly meditate and recollect. Even making notes seemed out of the question—so rushed were we. Therefore I felt a sense of agitation and despair about the memorabilia of this first day, as I entered my room at the hotel. How I longed then for all of you who are so close to me in the Spirit! The *multiplication of memory* is precious in our midst. I take real delight in the way each of you contributes to the composite revelation of the Lord in our midst. And your observations and impressions, as recorded in your Meeting Reports, letters and special writing, all fit together into a beautiful and rich tapestry, in which many trends of thought are intricately woven into a Master design.

Lying feet up on the wall, taking revitalizing breathing exercises, I earnestly prayed that the Lord would someway help me to do the seemingly impossible—absorb and retain all that He desired me to see and hear, and that I could just let the rest go. After ten minutes of this inversion and controlled breathing, I felt much better. (How thankful I was for the knowledge of good relaxing and revitalizing exercises that accomplish more in a few minutes than a lazy hour on a bed can do—for I never seemed to find a lazy hour!) Lunch, I decided, could be hurried through, and I might find time for some postcard writing. But I hadn't counted on the slow tempo of the waiters, nor the rush in the dining room. I ate lightly, resolving to keep my stomach shrunken, for we had been told that the less of the food and water we took, the less apt we were to get sick. When I could, I dashed back to the room and wrote a few cards. I also wanted to send a telegram, so everyone would know of our safe arrival. But when I got to the desk, I was told that the bus was ready to leave.

Chagrined to be late again, I rushed out, and found all the bus seats taken! Dwight was crowded in the rear between two huge colored sisters—and though he offered his seat, I couldn't even get to it. A brother put down a jump seat and told me I could crowd into it. (At no time did any of our brothers offer any of us sisters a seat on a bus!) I suddenly felt tempted to cry. It was so very, very hot—over 100 degrees in Jerusalem! And I was wildly thirsty, having not drunk enough water. I was also

tempted to be hurt that a seat hadn't been saved for me, and that they gave us such
uncomfortable crowded buses. The thought of taking that torturous ride down the
Jericho Road on this little narrow, hard, almost backless jump seat seemed really
overwhelming! I was drenched with perspiration and all the sore spots on my skin
were smarting and itching. I was just on the verge of feeling sorry for myself—which
is always deadly—when the Spirit reminded me that a real pilgrimage is not easy
on the flesh, but arduous. He had forewarned me about it, and I had agreed to come
and *be* His praise "throughout the Holy Land." After all, Jesus and the disciples had
walked this hot, tormenting road many a time. After all, riding a bus is better than
walking or riding a camel or an ass! So I lifted my heart to the Lord in thanksgiving
and depended on Him for grace for the journey. All around me there was a din of
fussing going on again. And in the heat and discomfort it seemed almost unbearable.
We were waiting for a missing brother—and nearly everyone was impatient. Our
guide was quite willing to continue waiting. But the group made such a fuss that he
finally told the driver to proceed. (This brother had to take a taxi to catch us, and
was charged $7.00 for only a few miles.) I could see that the guide thought we were
not very considerate of our brother, but he held his peace. He began to describe what
we were seeing as we turned down the Jericho Road. But I could hear little, for every
window in the bus was open, and the noise of rushing air drowned out other sounds.
So I quietly praised and even managed to hum a little, though I was beginning to
feel sick from the constant tossing about on the seat, since I had nothing to hold to.
I wondered how I'd ever make it to the Jordan! But suddenly the bus stopped and the
guide announced that we were going to pay a hurried visit to Bethany.

Bethany! Is there a place in the Bible any dearer? For more years than I can
recall I have cherished precious dealings of the Spirit about this blessed place where
Jesus found love and hospitality. First He taught me about Bethany—Faith; then
about Bethany—Love—sitting at His feet, the "Mary-heart" was revealed. Later,
the sacrifice of the broken alabaster jar and the costly myrrh—shedding fragrance
throughout the house—was reenacted in our midst in various ways. Again and again
I "visited" beautiful Bethany. And I know I speak for all of you in this regard—for
Bethany has been brought nigh a thousand times, and is indeed never very far "over
the hill." I had hoped to enter it physically by crossing over the Mount of Olives and

finding it just on the other side. Sitting in the bus, looking out on a few Arab houses at the bottom of a dry, weed covered hill, I felt a wave of revulsion. This just couldn't be Bethany! If it was, I didn't want to see it! (Later I learned that this village is now called *El Azariyeh*—village of Lazarus. The *ancient Bethany was located higher on the hill*—and more as we have pictured it.)

Since we must all hurry, everyone did. And in the press of the crowd, it grew very hot in the bus. Not a breeze was stirring! I knew I had best sit still, for I saw the guide leading the others up a steep hill. Another bus, with the other half of our divided party, had already arrived ahead of us—so there was quite a procession on the narrow path. I closed my eyes and looked to the Lord. Suddenly the Spirit stirred me. We were at Bethany! And this would be my first—and only—chance to visit it! What was I doing sitting there like a dope! I rose up at once, though sister body protested, and started up the path. It was wonderful to be alone, and to be able to talk to the Lord as I walked. Yet how long, steep, hot and dusty that climb to Lazarus' tomb seemed! I felt as though I had made quite a pilgrimage. Once there, I did not feel led to join the crowded procession going down into the tomb, but to retrace my steps to the little church of St. Lazarus which was off to one side. I recalled that it was one of the famous Barluzzi churches, designed by an ardent Franciscan lay brother who had devoted his life to designing and building new churches at important Holy Land shrines. He had sought divine inspiration and each is a unique "worship in stone." When I entered it alone, with a sense of awe, it seemed that it was a glorified tomb for Lazarus. It stands on the foundations of three older churches, the first built in the fourth century, commemorating the miracle of his resurrection. Portions of these churches remain in the apse, pavement or courtyard. On its gray walls I saw beautiful large mosaic lunettes and mosaic bands, with Bible texts in Latin explaining the life of Mary, Martha and Lazarus, recorded in the Gospels. The resurrection is depicted in a glorious manner in the cupola which is all of gold. It has a large open eye in the center, surrounded by 48 panels on which are most beautiful white doves, winging their way toward heaven, surrounded by flames and flowers. There are two simple, but elegant altars, in the form of sarcophagi. One has sculptured medallions of Mary and Martha. The main altar has two magnificent angels, pointing out the empty tomb of Lazarus. I was particularly impressed by the lunette which depicted Jesus at the supper given by Simon the Leper. The church is supposedly located near where his house stood. The doves drew my eyes and my heart again and again. And *nowhere else on our trip did I see such a glorious company of them*. They reminded me of you, my darling sister-doves, and of the beautiful "Dove Company" of which our Lord has often spoken.

On the outside I visited the ruins of a very ancient house which is said to be like the house of Simon, located at or near this spot. It was easy to picture Mary sitting at Jesus' feet in just such a room. By this time the others had caught up with me. I thanked the Lord for having let me go into the church alone and drink to the full of its message. It was quite evident already that many of our company had no interest in Catholic churches—and intended to deprecate them vocally, no matter who might be listening. How thankful I was—and am—that the Lord has taught us that the seven fine arts, of which architecture is one, were originally given by God to man to be used in worship. Beautiful places of worship speak to us, and we respond to them even though we know that our Lord dwells in temples not made with hands.

We were hustled back to the buses, and it was a relief to find that some of our company had taken seats on the other bus, which was not so crowded. I had a real seat now, and after sitting on the jump seat it seemed almost comfortable. I felt very drowsy and perhaps it was merciful that I began to doze, for the road serpentines almost as much as the Jordan River. And the atmospheric pressure steadily increases as the elevation lowers. So does the heat! It was by now about 108°. It is said that the pressure increases about 2.5 tons by the time one reaches the Dead Sea. The weight of this tends to make some feel sick, deaf or very sleepy. I felt a little of each. Vaguely I heard something about passing *Ain El Haud*, "The Fountain of the Apostles." This spring, the only one between Jordan and Bethany, is certain to have been visited by Jesus and the disciples. How I would love to have stopped and drunk from it—or from anything, for that matter—I was desperately thirsty! The further we went the more desolate the road became. But I caught only an eyewink view now and then. Soon we passed the many black caves and the huge rocks, where robbers used to hide. The next thing I knew we were coming to a bumpy halt at *Makkadet el Hajla*, "The Bathing Place of Pilgrims." This, of course, is supposed to be at or near the place where Jesus was baptized by John. It is located about five miles north of the Dead

Sea, and is more than a thousand feet below sea level. It felt like it! I stumbled out of the bus, still half asleep, and stared at the yellowish river flowing placidly before us. Its very steep banks were bordered with tamarisks, willows, acacia and the reed of Jordan—a kind of flowering bamboo. It looked like pictures I have seen—but seemed most unreal. Yet how very real the Jordan has been to me in the Spirit! The actual going proved quite empty in comparison with my "visits" to it. I recalled a snatch of the song which meant so much to me last spring: "I will journey with Jesus to the Jordan, and watch its sacred waters flow. And I'll picture the Holy Dove descending, as He did on my Savior long ago . . . ." Standing there, that memorable day, I felt nothing but burning thirst—I would gladly have drunk that water, could I have gotten to it. Instead we all turned aside to the inevitable cold drink stand which was doing a lively business. I can't tell you how odd and out of place things appeared to be: the drink stand, the tables for picnics, the souvenir counter with bottles of Jordan water and other tokens for sale! In addition to our two busloads there were other tourists there also. Someway it seemed incredible that the voices of John the Baptist, Jesus—and even of the Father—had once resounded here; or that here, as is believed, the Children of Israel crossed over into the promised land! It is said that on three occasions the Jordan parted here—for the Israelites, for Elijah, when he smote it, and for Elisha when he returned from Mt. Nebo. It is also believed that Ruth and Naomi crossed here when they journeyed from Moab to Bethlehem. I felt none of these stirring events as I stood on the bank that day and watched some of our party rent a boat to have their picture taken in it. I turned and walked away, as one in a daze. My eyes were hot and dry—though for some reason I wanted to cry. And my throat burned too—still thirsty.

Our next stop was only about five miles away—the world famous Dead Sea—which is neither dead nor a sea, nor is it ever called in the Bible by this name which all the world now uses. This lake has various names in the Bible. I couldn't recall them that day, but our guide reminded me of the one the Arabs use—*Bahr Lut*—Sea of Lot. In Gen. 14:3 it is called the Salt Sea; in Deuteronomy it is Sea of the Arabah or Sea of the Plain. Joel refers to it as the East Sea; in 2 Esdras and the Talmud it is the Sodomitish Sea, or Sea of Salt. Josepheus named it The Asphaltic Lake.

This body of water, often described as being dark, viscous, murky and malevolent in appearance, sparkled most beautifully blue-green and bright before my eyes! What a surprise! Blue-green is one of my favorite colors, and it is said to mean "wellbeing." To find the Dead Sea such a gorgeous color made me rejoice. And as I write to you about it now, the Spirit whispers that for those of us who have crossed Jordan, death has no darkness, terror or murky depths. It has become for us the way into LIFE—a life as sparkling, dancing and inviting as the waters that lay before me. The cobalt sky and the golden mountains of Moab on the opposite shore seemed to set it off to excellent advantage. It closely resembled our own Salton Sea.

We came to a halt at a rather modern looking hotel. And everyone felt somewhat excited to be at "the lowest spot on earth." I had wondered if I were to bathe and be literally baptized in that sea, or just dip my hand in it and be sprinkled. But I knew the Lord desired some outward sign. A few of the men headed for the bathhouses where suits could be rented. There were umbrellas along the strand, and people were playing about in the water or sitting on the sand. On the terrace food and drinks were being served. It appeared much like a California resort, except that the crowd was not large. We had another cold drink, which did little to quench our thirst, and I walked alone to the edge of the water. The greatest mineral wealth in the world lies in this sea—estimated in terms of between one and two thousand billions! Yet nothing living can exist in it, due to the concentration of salts. "This is the final depository of the Jordan, this is indeed the sea of death," I thought. And I recalled how my Sunday School teacher had said that because this water has no outlet, but attempts to horde all that flows into it, it becomes a place of death: only the stream that flows and gives can remain fresh and alive. This simple lesson returned from the distant past as I dipped my hand into the water. It felt oily, unpleasant—almost thick. And my hand turned red and began to smart. I placed a little on my forehead and it too burned. I noticed that none of the bathers stayed in long, though they enjoyed floating around in it—since it is impossible to sink in such buoyant water. (Later they said that their eyes and nose burned for hours.) How significant it seemed to me that the sea of death holds one in buoyant embrace and claims no human lives! It is God's will that all men be thus upheld and passed from

129

death to life. Salt is a symbol of grace. As the "salt" upholds the swimmer, so the grace of Jesus Christ upholds the believer.

I meditated a little about the Dead Sea Scrolls and other ancient manuscripts which were found not far from this spot. And I thought again of Sodom and Gomorrah, said to be buried in the south end of the sea. Most of all I wondered about Ezekiel's and Zechariah's strange prophecies about this sea becoming eventually alive again, and sweetened with living waters. Well, the Lord sweetened the waters of Marah—and could surely transform a sea as well!

On our way back we were to stop at Jericho, which means fragrance, palm trees. I had long anticipated visiting this city, one of the oldest in the world. It has always fascinated me—a lovely oasis in the midst of such desolation! The modern Jericho is built on the site of Byzantine and Crusader Jericho and has many beautiful gardens, groves and orchards. We did not stop there but continued on to Elisha's Fountain—which is still flowing. (2 Kings 2:11-13; 25:2-7) Extensive excavations are being carried on in this area, and also on the site of Joshua's Jericho—about a mile to the north. We were taken to one large tell (a hill formed of the debris from the past), and shown the various layers—each depicting a different period of time. So far, the archaeologists believe they have reached back at least 7,000 years! It is interesting to us that the walls that fell under Joshua's leadership have been partially excavated! As I stood there amid such antiquity and watched the native Arab women and children drawing water from this ancient spring, I was deeply impressed. I had longed to walk in the Jericho Jesus knew and loved. And I was still confused about just where *that* was. I left feeling that I must surely pass this way again—if not on this trip, then, please God, at some later time. I have a tryst to keep in Jericho!

Standing nearby is the famous Mt. of Temptation—called Quarantine, 40 days—on which Jesus is said to have fasted and prayed. There are many rocks and caves on this rugged mountain. Today a Greek Monastery and the ruins of a Crusader Church remain. It is remarkable that the ground on which Jesus gained His victory over the devil is close to the spot where Joshua gained his first victory over the Canaanites! I marveled again that last spring the Spirit led us to a similar mountain near our desert, and that it was more truly made the "Mt. of Temptation" to me than this ancient one on which I now gazed.

Jericho and the famous mount looked especially glorious as we left them behind, journeying once more toward Jerusalem. The afternoon shadows were falling. The blasting heat had tempered a little. A hazy veil seemed to lie over this ancient place—which holds in its bosom many hidden secrets yet to be discovered. Again we were on the Jericho Road, going up to Jerusalem. Our last stop was at the Good Samaritan Inn, now a Police Station. Winding on and on, up and up we went, hoping for a fine sunset view of the city! Again the road seemed unduly long and wearisome. And evening shadows were upon Jerusalem when we whirled again around the bend that turned us toward its walls. Our first day of touring had ended. An amazing day! And it was no wonder that we were tired. We had walked outside its walls in the early morning, had toured within its walls from midmorning until after the noonday, and had traversed the Jericho road three times in less than 24 hours!

O LITTLE TOWN . . .

I n the midst of the din of Jordanian traffic and the clatter of our companions, I heard a word that aroused me from my meditations like the sudden pealing of a bell. "Bethlehem!" Oh beautiful word! O beautiful little town! I can think of no more evocative or nostalgic place in the world.

> To an open house in the evening
> Home shall men come,
> To an older place than Eden
> And a taller town than Rome . . .
> To a place where God was homeless
> And all men are at home.

—G.K. Chesterton

Dear Bethlehem! Again and again we have made the pilgrimage to Bethlehem in the Spirit. Song after song has centered around it. The manger has been become the heart of "the heart of the world." And of course there was a special reason for me to hasten to Bethlehem—for our little Arab "David" boy lives there. And the Lord was sending me to see him and give him our love.

It was our guide who had spoken this magnetic word. But what he had said was lost in the hubbub. I could hardly wait until the bus drew up to the hotel and I had a chance to get to him and inquire about it. "Tomorrow morning, bright and early, we go to Bethlehem," he told me. "Oh praise the Lord!" I responded, and almost bounded off the bus. All the weariness of the long, hot journey momentarily fell away, and I was bubbling over with praise when we entered the lobby. Dwight was already asking for our mail, and when he handed me a nice stack of letters, my cup really ran over. (You dear, faithful darlings!) I pictured myself on my bed, feet up on the wall, having about an hour's feast with the mail while I rested. So I hurried to

135

our room. But things never seemed to go quite like I pictured them! When I opened the door I had another surprise! This time it was a pleasant one: in the middle of the little room sat Dwight's no-longer-missing-bag! I went back to find him and tell him, since he was waiting in line to make further inquiry about it. With thankful hearts we set about unpacking our things and hanging them. Now, at last, I could curl my hair and get rid of my frowzy look. The Lord had given me grace about it, and I had been quite at peace all day, even though I realized how disheveled it was. But it *had* been a trial to me that nearly every time I turned around—and sometimes even before I quite made it—my "camera-happy" husband had caught and preserved my tousled head and perspiring face.

Having my picture taken is always a trial to me, even when I am reasonably prepared for the occasion. I had faced all this before coming, and had settled it that I would cooperate fully with Dwight and be as pleasant and willing a model as I could be for "the duration." I knew that his pictures would be made into slides and be shown again and again. Though I shrank from being thus enlarged and projected on a screen over and over, I just had to accept it gracefully. I realized that I would be hot and dripping with perspiration most of the time; my clothes would inevitably get wilted and wrinkled; the wind would blow and muss me up; and I would get very dirty and weary. All this I had counted on and was prepared to accept. But I had not foreseen having to do so without some sort of hairdo! So I was thanking the Lord, as I unpacked, that this test was over. One day of it had been quite enough! Or was it? At about this point in my thanksgiving I reached the bottom of the bag—and still no curlers! Oh no! (The curlers had just disappeared! We never did locate them at home either!) Now I would have to visit The Christian Approach Mission and meet all those people, for the first time, looking frumpy. And what about Brother and Sister Mattar and their associates? Surely we would have to see them also on the morrow—and visit the Mount of Olives chapel! It is strange how such little things can seem so large at our weak moments. I collapsed on the bed, trying hard not to feel depressed. Then I remembered the letters! I borrowed Dwight's reading glasses and after a few minutes with the letters I felt as though I had received a real transfusion. So I decided to be valiant—curlers or no, I would someway carry on! There was not time to read all these interesting epistles, for the dinner hour was approaching. We needed our baths badly, so we began preparing. A few moments of breathing exercises and stretchings gave me a real pickup. But oh, how I would have enjoyed a nap!

Dinner was a very pleasant and leisurely occasion. And the food was really very good. It had been a little problem for us to know just where to sit in the dining room. At breakfast we had sat at a long table where a number of our party were eating. I guess they were special friends, for they seemed a little odd about our sitting with them—and directed their conversation to each other rather pointedly. At lunch it had been somewhat the same. I was surprised to find this attitude among such a group, and I realized that perhaps those who were traveling alone felt it too. We decided to take a small table and to sit with men or women who appeared lonely. On this night we selected two middle-aged women who seemed really glad to be with us. One of them mentioned how much she had been uplifted by my singing in the Spirit on the previous night. I was startled; it seemed it must have been a week ago! Was it possible we had not yet been in Jerusalem even 24 hours! We had opportunity to talk about the Lord, His goodness and faithfulness, and of our precious privileges in visiting the Land. I felt a real anointing to testify and praise Him, and our sisters seemed to respond. They each spoke of how disappointed they were in the unspiritual attitudes of the group.

After dinner I felt moved upon to try to contact Brother Mattar, and also Souad, our little adopted blind daughter in Jerusalem. I realized that time was going fast, and our stay in Jordan would be all too short. I also wanted to send a telegram to the Mount, telling of our safe arrival and prospective journey to Bethlehem. But I learned that the Telegraph Office had closed and would remain so on Sunday. Only a message required by an emergency could be sent. We had already been told that in Jordan there are two holidays a week—Friday is the Moslem day, and Sunday is for the Christians. Schools and some public offices are closed both days. Repeated phone calls failed to reach Bro. Mattar, and the desk clerk seemed blank about St. George's School, where Souad lived. I began to feel frustrated. I was looking to the Lord, but our time was growing short fast. I knew for certain that we *must* see the Mattars and the Chapel. Seeing Souad was only a desire—one which I hoped the Lord would grant. However, just before I left for the trip, I had received a letter from the

Christian Children's Fund, saying that she had been recently married (at fourteen!) So I knew she had likely left the school, though I hoped to trace her through it.

By this time it was nearing nine o'clock. I still had letters to read and also wanted to write some cards. But most of our people were out shopping, and I suddenly thought again about the curlers. Maybe I could locate some bobby-pins at least! Out we went with the crowd. But the only shops open were those displaying gifts and souvenirs. We decided to pick up a few, while we had the chance. And the time sped by. Right in the middle of my shopping I suddenly remembered my washing! A good traveler keeps her personal washing promptly done up. But both Dwight and I had forgotten about this in the rush. Reluctantly we returned to our dimly lighted room and collected a rather large hand washing. My nice stretchy clothesline provided all the hanging room needed. And my little hand-size plastic washboard was a real help. Yet it took time. We were both very weary, and could hardly see to watch, let alone to read by such a light. So I put away all thoughts of writing cards, reading the rest of the mail, and jotting down notes. Our first long, wonderful, exciting day on tour had ended. My mind had taken thousands of pictures. My little "recording machine" was jammed! And I began to feel depressed because so much I had heard and seen was already dim or had blended into a confusing montage.

We slept soundly during the almost holy silence of that Jerusalem night. All traffic and noise seems to subside early, and a hush settles down over the city, making it seem rather awesome after the constant clamor of its teeming life by day. Then, shortly before the dawn, the city wakes up and begins running about and shouting like an eager child—and there's no more rest or quiet. On this, our second morning, I had no desire to go out into the streets, though Dwight did. By now I longed for a little time alone with the Lord in prayer. It passed all too quickly. My heart was still the child, but my body felt very "grandmotherish" on this memorable morning. It hurt and ached from head to toe. Nevertheless I dressed with care for my pilgrimage to Bethlehem. I know Christmas is the desirable time to go, for then the holy processions are formed, and with singing and praise and much joy pilgrims from all over the world go into the city. The bells are rung; the Shepherd's Field is illuminated; services of adoration are held throughout the night. How different it all seems in May, with blue skies over it and the golden harvest fields all around!

However, since I could not go at Christmas time, it seemed to me that Mother's Day was the next most blessed time to go. I felt the wonder of it as I prepared, and I recalled that one had written to me about the pleasure our Lord would have in taking me to His birthplace on or near Mother's Day.

I was almost too excited to eat breakfast, but I tried to have presence of mind and radiate love to all in our own group and to those who served us. A phrase from a little song that often comes to us was running through my mind: "And my heart starts to beat, like a child when a birthday is near. When you are in love, it's the loveliest day of the year." I must confess that I thought little about my natural family on this Mother's Day, and not too much about my Spiritual Family. We had already had our Mother's Day, before I left. This day seemed to belong to Jesus and to Bethlehem and Jerusalem, in a special way. And it was a day I hope always to cherish.

The previous morning I had put on my large, lightweight sunhat, which I had been led to prepare for the trip. And oh, how it was needed! The sun almost "smites" one over there, as Psalm 121 mentions. But it did not seem appropriate to wear this to the Mission, so I wore a neat little veil-hat instead. I rather hoped we would get there in time for church—I so longed to attend service while in the Land. I was soon wishing for my hat, however, for our bus seemed in no hurry to get to Bethlehem, after all, but headed instead up the Mount of Olives. Even to mention this mountain stirs my heart! So you may well imagine how thrilled I was to ascend it to the very top. The Government has provided a little park-like place up there, and there is a sort of tower and platform from which the entire city and environs may be viewed. I walked slowly along the roadway, knowing that Jesus walked there too, or nearby. It was a little shock to see a barbed wire-encircled camp of soldiers. They, along with a few Arab families are the inhabitants of that Holy Mountain—in addition, of course, to the various churches on it. I tried to picture the great Roman Army of Titus camped there, as it was in 70 A.D. What a contrast to this handful of Jordanians! Babylon hordes too had pitched their tents here. In fact army after army, encircling and making siege of the Holy City, chose this mountain as their main campsite. Yet it is called the Mount of Olives—the symbol of Peace!

It was happily peaceful that bright morning, but not very quiet. Tourists were all over the little park. And we had to await our turn to enter "The Ascension

Chapel" or mosque, which is supposed to market the place where Jesus last stood just before He was taken up. Of course there is no way for us to be sure about this, for the Word also says that He led the disciples out as far as Bethany and blessed them—implying that Jesus ascended near Bethany. But in another place we read that this occurred only a Sabbath Day's journey from the city, or about 1,000 yards. This is just about the distance from Jerusalem to this Ascension Mosque. In 378 the Roman lady Pomenia built a polygonal building here. Its form was followed somewhat when the Crusaders reconstructed it later, calling it the *Imbomon* (summit.) It was octagonal and had a concentric row of columns, (the bases of which may still be seen), and carried a circular "drum" surmounted by a cupola open at the top. In the center is a stone where the footprint of Jesus—according to legend—is imprinted. In 1187 this little building was turned into a mosque, and it was again reconstructed in 1835. Although it belongs to the Moslems, there are altars on the outside for the Armenian, Greek, Coptic and Syrian churches.

Our Moslem guide seemed to take special delight in taking us into it, showing us the large footprint, which he said was of the right foot of Jesus, and telling about His second coming. How very strange it seemed to me to hear this glorious message from a Moslem! I can still see the happy glow in his eyes as he knelt reverently by the stone and spoke of Jesus. One of our brothers asked him if he really believed what he was saying. "But of course!" he answered, "Jesus will come again." The brother then said, "You must be a Christian." "Oh no," he answered, "I am Moslem." I think he noticed that most of us were unimpressed by the rock—even though it most certainly did have an indentation like a large foot in it. We were all much more concerned about when Jesus' feet shall *again* stand on this mount, perhaps at or near this very spot. And the glorious hope was freely voiced among us.

We climbed up on the roof to view the city, trying to escape the souvenir peddlers. But it was too crowded for comfort. It seemed that everyone was either taking or posing for a picture. Again I felt near tears. Here was the blessed Holy City both the Jordan and the Israeli sides—in full view! My heart was swept with a surging emotion, much as a harp swept by the hand of the artist. I wanted to praise, laugh, cry and sing all at once. But I just stood there silently. To my left a small group were withdrawn in a little circle. Unconsciously I edged toward them. Among them I heard a voice that touched me deeply, and soon I caught a glimpse of the speaker—a young man, slight of build, with dovelike eyes and a gentle voice. He was pointing out the various buildings in the city, and kept speaking of our Lord in the most loving and gracious way. Then I heard him say, "Now look over there to that tower, and behold Mount Zion!" It was almost as though Jesus Himself had spoken directly to me, and was showing me His City. Oh how I longed to cross over into Israel and climb that sacred hill too! I realized with a pain that soon we would be on our way to Egypt, leaving all this behind. And my heart was very sad. I lingered with this group as long as I dared. It was such a joy to hear a Christian guide telling about the city. And some way that young man sweetly lived on in my memory.

By the time we were back in the bus, we were very hot and thirsty. And most of us were restless to be off to Bethlehem. But our next stop was at the foot of the Mount—at dear Gethsemane. Of course I had hoped to come here by night, for the first time, instead of in the full glare of a bright day. It seemed most unreal to me. Since this was a Sunday, tourists were more plentiful than usual and the entrance and garden were jammed with the ubiquitous cameramen and women. No matter where I stood, someone would be sure to ask me to move, as I was spoiling their view. There seemed to be no sense of reverence or worship, but just a curiosity about how old the trees were, and was this or wasn't it the right spot?

I got away from them and entered the Church of the Nations—"The Basilica of the Agony"—as soon as I could. I have read about this beautiful and inspiring church, designed by Barluzzi and built by the Franciscans—who have had possession of Gethsemane since 1681. And as I entered it the Spirit moved on me strongly. I walked about with uplifted hands worshiping quietly for some time. The Spirit made me oblivious of others, and I'm sure I did not disturb them.

It is said that the first church to be erected here in the Garden of Gethsemane was built in 380. And there are many who believe that this is truly the place where Jesus prayed on that memorable night before His passion. At that time, however, almost the entire mountain was covered with olive trees, and some believe He was somewhat higher up. The present church was built in 1925 and is most magnificent. It surrounds the "Rock of the Agony"—a large rugged rock which was most certainly there in Jesus' time, and may indeed be the very rock He used for an "altar" as He agonized. It was very impressive to look upon this rock, behind which the church altar has been erected. How often the Spirit has led us to find a rock altar as we worship in the mountains or at the sea! I thought of our rock sanctuary and our many altar stones. And I longed to cast myself upon this rock. "He that falleth upon the rock shall be broken . . ." But of course it is carefully protected by the altar railing. Off to one side I found more large rocks, left just as they have stood for centuries, the church being skillfully built around them. There are also some mosaics which are said to have been preserved from the first church. All the windows are of vari-hued alabaster. The lighting is therefore dim, and almost funereal. Yet, after I became accustomed to it, I felt exalted into another world—a glorious Kingdom realm where heavenly light was dawning. Six columns support the twelve cupolas, which give the exterior of the church an unusual grace and beauty. The effect Barluzzi desired was that of prostration before "The Rock of the Agony"—not of space or height. And he has beautifully achieved it. This is indeed a "worship" in stone and alabaster.

Feeling transported into the Kingdom, I paused a moment under the coat-of-arms of each of the nations that helped to build this church. Never did America's eagle appear more glorious! How proud I was that we were displaying our colors here in mosaic, in honor of the agony of Jesus Christ—the "desire of nations." I prayed for each nation, passing from one to another, remembering our special dealings concerning them regarding the latter day: Germany, Canada, Belgium, England, Spain, France, Italy, Mexico, Chile, Brazil, Argentina, Poland, Hungary, Eire and Australia. I noted that Russia was conspicuously missing from the list even back there in 1925!

I could not help but feel, however, how very incongruous it was to have the coat-of-arms of the great nations displayed here, when Jesus desires all men to lay down their arms and enter into the peace of the cross. But that day will come in the Kingdom!

I might mention here that I had greatly admired the facade of this basilica every time we passed it on the Jericho Road. It is a glorious mosaic representation of Christ offering up to His Father His sufferings and those of the world. Its gold background makes Jesus stand out in special beauty. There are four large Corinthian columns placed between three large arches. Above the columns are statues of the

Four Evangelists: Matthew, Mark, Luke and John. And inscribed above the portal are these famous words: "Tarry ye here and watch with Me."

These few minutes spent in free—though discreet—worship were most reviving to my spirits. My little journey into the Kingdom had to end, however, when I saw that the last remaining ones of our own tour were leaving the church. I followed quickly. In the little entrance hall some were buying slides and souvenirs. I started to pick up a few things and was accosted by one of the ministers of our group. He spoke most crudely about the Catholics and the Pope, and told me not to buy one thing, or I would be supporting the Pope. Others were clamoring about the Roman Catholics having these Holy Places, and saying derogatory things. Here we were, the guests of the gentle Franciscans, in their house of worship, acting uncouth and ungrateful! I reproved this brother gently and told him I thought the Pope would likely fare well, even without any help from us. And I said, "After all, if these Franciscans and other Church Orders had not sought these Holy Places and obtained them, they would long ago have disappeared or been destroyed. I am grateful to all who have preserved them, even if they have built over them and changed them somewhat." But of course they paid little attention to me. Dwight suggested that we return to the garden on our own time later (which we did), so I shall write in more detail about the Garden in that portion.

At last we were all reloaded on to the bus and were jogging down the Jericho road again. But we turned off at *Ras el Amud*, and passed through a little valley. We began climbing quite a steep hill, following a winding mountain road. Then down again we raced into another valley and up another hill. Each turn afforded a delightful, though somewhat similar, pastoral scene. We saw the harvest being reaped by primitive methods; sheep being tended by biblical-appearing shepherds; and the Arabs dwelling in their simple houses, surrounded by children and animals, much as they have for thousands of years. So thrilled were we over the "Sunday School paper" pictures all along the way, that we scarcely noticed how many and how sharp the turns were. From time to time though, we let our driver know that we were a little jumpy, particularly when we whirled within an inch or so of other cars on a turn. But somehow we survived the seventy-five or so sharp curves which had to be negotiated between Jerusalem and Bethlehem. I am sure you all know that the reason we had to take this winding road is because the old, direct route to the "Little Town" is now partly in Israeli territory, and has had to be closed because the two nations have no dealings with each other. This means that although the distance is only five miles, we had to travel twelve instead. And twelve miles over that kind of road is about like thirty or forty in California. Living in Idyllwild has accustomed us to curves and mountain roads. But the only way I could stay calm was to watch the scenery and not the road!

There was a high plateau reached at one point, where we could catch a glimpse of the Dead Sea and the Mountains of Moab to the east. And again we saw over into Israel at various points, and once passed within a block of its border. So near, and yet so far! Our guide told us over and over, "There where it is green, is Israel." (Yes, he used that naughty word, though many Jordanians will not.) "And where it is dry, that is Jordan." Then someone would ask him why, and he would shrug. The Arab has never seemed to understand how to conserve his land and water. But Israeli "know how" has played a big part in transforming their once barren land into a garden. Yet the Arabs in Jordan go on in their ancient ways, and the result is a sadly desolate land in most areas.

We were all eager for our first glimpse of Bethlehem. And it was well worth waiting for! Suddenly we saw it—off at a distance, across a long valley—sitting serenely, as pretty as a picture, framed between two hills. The bus driver slowed down long enough for us to enjoy this view, and then proceeded to do a lot more upsies and downsies, turnings and twistings. Soon the new road had joined the old road (the portion on the Jordan side) and we had the joy of knowing that we were about to enter Bethlehem on the same road that Joseph and Mary—and all other biblical pilgrims traversed. We passed the *Mar Elyas* Greek monastery which some believe marks the place where Elijah rested when fleeing from Queen Jezebel. Then we had a closer view of "the Little Town"—"seen across a fertile valley like a toy town of white blocks, built on the slopes of a green terraced hill, surrounded by rugged barren mountains. This district has always been noted for its fruitfulness, its fine fields of grain, vineyards, orchards and flowers. Its very name, of course, means House of Bread—"*Beth Lekhem*" in Hebrew; "*Beit Lahm*" in Arabic. And it is said that primeval wheat, the original of all cultivated species, has been found wild there. It was truly a fitting place for The Bread of Heaven to be born.

We traveled the same road the Kings of the East followed on their way to the Baby Jesus, and just outside the little town we came to Rachel's Tomb. Our bus stopped so we could visit it. Both Jews and Moslems have honored it through the centuries. Now, of course, it is in the hands of the latter. The present "tomb" dates back only to the 15th century, and is an oblong stone building with a whitewashed dome. The Spirit has made Rachel and her two sons very dear to my heart, and it gave me a sense of reverence to remember this mother in Israel on Mother's Day (Gen. 35:16-20)

Once more back in the bus, I caught my breath a little and instinctively put my hand over my fluttering heart. The great moment had come—and I was about to "come home" to Bethlehem, beloved "little town!" Micah's golden words were ringing like a bell within me: "But thou, Bethlehem Ephratah, though thou be little among the thousands of Judah, yet out of thee shall He come forth unto Me that is to be ruler in Israel; whose goings forth have been from of old, from everlasting."

I wanted to sing the song we so often use at Christmas, given by the Spirit to the melody, "Memories."

> Bethlehem! Bethlehem! humble little town,
> The King of Heaven has favored thee with glory and
>
> renown.
> Bethlehem! Bethlehem! All the nations of the earth
> Shall come afar, still following thy star
> To the place of our Savior's birth.

THE CITY OF DAVID

A   nd Joseph also went ... unto the City of David, which is called Bethlehem; (because he was one of the house and lineage of David;) to be taxed with Mary, his espoused wife ...."

Who could make a journey to "The City of David" without thinking of him along the way? Our blessed Lord lived there only a short time during His infancy; but David spent all his childhood and young manhood in this area. In the hills and fields surrounding it, he lovingly tended—and defended—his flock. His beautiful voice was often lifted up in song and prayer, and in reciting aloud the Torah. He spent long lonely nights under the stars, and quiet days among the fields and flowers. He became intimately acquainted with nature, here in this charming setting. And when he later sang about the sun, moon and stars, the wind, the rain, the lightning and the thunder; he voiced, no doubt, many of the concepts that came to him during those golden days of his youth. It was said in Israel that David often heard the songs of the angels, and that he "stole" them and translated them into earthly language. It is quite possible that in the very fields where the shepherds saw the angels and heard their chorus, on the night of Jesus' birth, David too heard angel voices and saw scenes of heavenly splendor.

Then came the suspicious day when the great prophet Samuel arrived in Bethlehem, in obedience to the command of the Lord! At the home of Jesse he prepared a sacrificial feast and told them to sanctify themselves unto the Lord. For Samuel was appointed to anoint one of them as God's chosen King over Israel. One by one Jesse's seven sons were scrutinized as they passed before Samuel. But all were rejected. Meanwhile, David, the youngest, was out keeping the sheep. "And Samuel said unto Jesse, send and fetch him . . . . and he sent, and brought him in. Now David was ruddy, and withal of a beautiful countenance, and goodly to look to. And the Lord said, Arise, anoint him: for this is he. Then Samuel took the horn of oil, and anointed him in the midst of his brethren: and the Spirit of the Lord came upon David from that day forward."

I had a painfully sweet desire to commune with the Lord—and with David—in Bethlehem. I wanted to walk in those fields, climb those hills, and sit on some of those ancient rocks. (How often David sang of the Lord as a Rock!) I longed to sing some of the lovely Davidic songs the Spirit has given to us. I recalled that I had read that there are no memorials of David in Bethlehem, no clearly identified places where he lived, and no altars erected in his memory! And this had struck me as being very odd indeed—for the Holy Land is teeming with altars and shrines, and it is said that even the Moslems love and honor David and quote from many of his psalms. It must be that David has been eclipsed by the rising of His brighter Son! However, even if there is no material reminder of David in Bethlehem, I am sure I am right in saying that something of his spirit lingers there even after 3,000 years! Here he was born! Just where, we cannot be positive. But many believe that his family lived on at least some of the property owned by Boaz and Ruth, his ancestors. This same property was still owned by some of the Davidic family when Joseph and Mary arrived centuries later, and *the Inn stood on it.* So there is a tradition that David's home was among the grottoes which are now beneath the Church of the Nativity. The only other spot positively identified with David is a well in the western part of the city. It is believed that when David's brave men stole through the Philistines lines and risked their lives to get David a drink of Bethlehem water, for which he thirsted, they drew it from this well. But darling David, swift to realize and appreciate the sacrifice they had made, refrained from slacking his thirst, and poured out the water as a drink offering unto the Lord!

Oh, what a desire I felt to honor our beloved David in his own little city, the place of his birth and early life! And I knew that each of you would have felt as I. What a blessed pilgrimage we could have made as a company, lingering in the little town and walking and worshiping among the hills! As my heart turned toward you, I was very thankful for each one of you and your precious "Mary" ways! Tears filled my eyes as my heart overflowed with a fervent desire that you might someday make this same pilgrimage in person.

Meanwhile our bus was racing along a narrow road that clings to a hillside. I wiped the tears from my eyes and watched closely, wondering what we would see first as we entered the actual town. To my surprise it was a familiar building—our own Christian Approach Mission! Spread along the road are five large stone buildings, each distinctly marked in English signs: the first was the Christian Approach Orphan's Home, where Awni and the other orphans live; next stood Boy's Dormitory No. 1; then Dormitory Annex; the Mission School; and the Crippled Children's Convalescent Home. These buildings overlook The Shepherd's Field and the Valley. Rounding another curve, I saw their Mission Trade School and the Church. I surely wanted to stop there, for it is said that one may get the best view of Bethlehem and the Church of the Nativity from this point. I felt a thrill of joy and thankfulness that the Lord has given the Mission such an eminent and beautiful location.

By now we were driving over the narrow cobbled streets of the little town, commenting about how clean, attractive and shining it appears. And so very biblical, thank the Lord! Almost all of the inhabitants of Bethlehem are Christians and proud of it! They are delighted to be living at the very Heart of Christendom, and enjoy the constant stream of pilgrims and tourists coming from all over the world. Many of them make their living by fashioning the beautifully carved mother-of-pearl ornaments and jewelry for which Bethlehem is famous. Others carve religious symbols and souvenir articles from wood. So of course, the streets are lined with little shops, or bazaars, as they are called among the Arabs. And on every hand the friendly Arabs smiled and bowed and welcomed us in their delightful manner. These Bethlehemites did not seem to mind being photographed, as do most of the Jordanians, so the cameras all got busy at once. My heart was flooded with love as I walked slowly from the bus toward Manger Square—the cobblestone-paved court in front of the beloved Church of the Nativity. I wished everyone would enter the church reverently, in silent worship. But my wishes were in vain. So I followed along to the entrance, which is a small low opening in a very thick wall. Even a short person must stoop to enter the building erected over the cave where Christ was born!

This cave is said to have been jealously preserved by the early Christians. And few doubt that it is authentic. It is known that the Romans deliberately planted groves for pagan worship on the site of each Christian shrine. It was Hadrian who

151

planted a grove here—around the cave—to the honor of Adonis. But in A.D. 330 it was cut down by order of Constantine, and a great basilica was erected over the cave, which his mother Queen Helena had discovered during her devout pilgrimage in the Holy Land. This Church was destroyed in 529 by the Samaritans, but some of its mosaics were preserved and may still be seen. The present Church was built in its place, so it is very ancient. I believe this was the only Christian Church to escape destruction during the Persian Invasion in 614. And there was an interesting reason: the coming of the Magi was beautifully depicted in mosaics near the entrance. When the Persians saw these they recognized the Magi as Persians, and spared the Church. Again in 1009 this blessed old Church of the Nativity faced destruction—this time from the Egyptians. But the Caliph Hakim, a devout Moslem, so respected Christ's birthplace that he too spared it. According to tradition, some of the columns of the church were used in Solomon's Temple. In any case it is a very ancient and venerable house of worship.

One by one we stepped through the little door and paused, trying to adjust to the dimly lighted interior. For outside the sunlight had been dazzling on the cream-colored stones! I lingered behind the others, wanting to lift my heart to the Lord before proceeding. As my eyes grew accustomed to the dim light, I was surprised to see a charming well-dressed Arab man walking toward me. He held out his hand and said, "Welcome to Bethlehem!" For just a moment I felt as though Jesus had come to meet me and taken me to His birthplace, and my heart skipped a few beats. Then I realized that he was saying, "I am Mike Handal, the secretary of The Christian Approach Mission." News evidently travels fast in the "Little Town!" Just how he knew we had arrived, I do not know, for I had not been able to tell him in advance which day we would visit, though we were expected at an approximate date. A flood of joy swept over me as I greeted him in Arabic. By this time Dwight was at my side and we were arranging with him an appointment at the close of our tour in the Church and the vicinity. Our guide also came to our side, wanting to hurry us along. He graciously agreed that the bus would take us to the mission.

When I finally turned to look upon the basilica, I was surprised that it was so large and impressive. Its simplicity is charming. Many of the churches in the Holy Land have a distinctly cluttered appearance—at least to those of us not accustomed to the formal Eastern churches—but not this one. The Basilica is 100 feet long and 70 feet wide. There are four rows of massive pillars—two on each side, marking double aisles. At the top of each pillar a cross is carved. Under the peeling yellow paint can be seen the beautiful native red limestone, which is marked with white streaks. On the walls are dim mosaics depicting the Nativity. Our guide led us down the nave and through the transept until we came to a stairway entrance to the grotto. Then we went down two flights of winding stairs and found ourselves in the cave in which Jesus was born. It is about forty feet long and twelve feet wide. Some thirty-two lamps illuminate the cave dimly. At the east end a large fourteen-pointed silver star is set into the floor in a special niche, around which richly colored velvet drapes are hung. The inscription is in Latin: *Hic de Virgine Maria Jesus Christus Natus Est.* "Here Jesus Christ was born of the Virgin Mary." It is said that the Crimean War began when someone stole the star, because of religious disputes about this holy place. (And the disputes are still going on—even on this Christmas, 1961, there were most unChristlike incidents here.) The Greek, Latin and Armenian churches share the grotto. And surrounding the Church are the walls of their convents. A Moslem guard is always posted at the birthplace of Jesus, in order to keep the Christians from fighting over it. What a travesty! On special holy days additional guards have to be added to keep the peace!

I found it dismaying to visit this ornate Holy Place—which certainly looks nothing like it did on the night when Jesus was born—with a crowd of chattering people. And it was an additional shock to discover a professional photographer standing ready to take pictures of each of us kneeling by the star under the ancient lamps! There was no opportunity at all to be alone or worship quietly. So I knelt quickly and worshiped in haste, so someone else could have their picture taken in a hurry. Dwight knelt with me and told them to snap a picture of us. And I agreed, realizing that he was in very few of our pictures, since he was the photographer. As I look on this picture now, I can see by the expression on my face how strained I felt and how unfamiliar it all seemed. I am so thankful that the Spirit can take us to the manger in a far more realistic way—and often does!

Three steps led us still further down into the "Chapel of the Manger." A marble manger now stands where the true manger is said to have been. (He was born several feet from the manger in which He was placed.) Some say that even the manger was formed of rock, cut out of the cave. Another famous cave is close by, and we found it by following a narrow passage—the cave of St. Jerome, where he lived for so long, making his Vulgate translation of the Bible. The Church of St. Catherine (Franciscan) adjoins the Basilica of the Nativity. We were given only a hasty peek into it, as we were hustled along. It is here that the famous midnight mass is conducted each Christmas eve. Perhaps you, as I, have heard the beautiful Bethlehem bells ringing at its close—for they are carried by radio all over the world. How I wished they might ring out freely that auspicious Mother's Day morning. But I heard only a few tolls.

Reliving this golden hour spent in Bethlehem, I realize that we were just rushed from place to place. There was no chance to walk in the fields and think of David and Ruth and other precious ancestors of Christ. There was no opportunity to sing, worship or pray in a quiet place. (Though I did manage to hide some of our tokens in special places along the way.) There was no mention of or opportunity to see the beautiful new Chapel of the Angels and the Shepherds, or even to visit Shepherd's Field, at which it is located. St. Jerome, in about 386, carefully established the location of this marvelous bit of earth over which the angels hovered and sang to the favored shepherds on that first Christmas night. This beautiful and worshipful chapel is the last work of Barluzzi, and was completed in 1954. It is said to perfectly express in architecture the memorable words recorded in St. Luke. It has an indescribably heavenly atmosphere, I have heard. And it is a *must* on my list, if and when I return to the darling little town.

I began to feel like a tired sheep being shoved from place to place. So I behaved like a goat and "got out of line." I ran as fast as I could into a little bazaar, hoping to escape the clamoring crowd. I was greeted by a dear Arab woman who was very motherly. We loved each other at sight! We embraced, praised God, and talked about Jesus—she in Arabic, I in English. I bought a Bethlehem woman's headdress from her, and she put it on me. As she did, I felt the Spirit come upon me in a sweet way, much like a mantle had been received. I knew it was sweet Jesus again, moving this time in this simple woman. However, I didn't want to leave the headdress on, for they are high and very conspicuous. But she begged me with tears to leave it on. And she taught me to say, *"Ana Beit Laham sitt."* (I am a Bethlehem woman.) She also taught me one or two phrases I wanted to use with Awni, and wasn't sure of. We parted with more embraces, and she begged me to write to her. In another little shop I found a white and red native dress, and Dwight appeared about that time and urged me to get it. I barely had time to get my package before the guide appeared. As we stepped out of the store, this guide, whose name was George, gently took my arm and tenderly led me down the rough cobbled street toward the bus. At first I was inclined to be embarrassed, but suddenly I felt Jesus in him. We had truly been praying for him and Ahman, and at times George seemed to melt as we spoke of the Lord. I had a strange feeling that Jesus Himself was walking with me by the sacred place of His birth and on toward the bus. We must have made a very strange picture indeed, this young Arab man in Western clothes and I in a Bethlehem headdress and veil. But on we went, with the natives and our tour—companions alike, laughing and making merry. But no one minds a little fun over there.

In less than five minutes we were at the Christian Approach Mission. Our guide urged us to get back to Jerusalem in time for the afternoon tour, and said they would wait for us. So we got off amid the well wishes of our companions. Most of them knew about Awni by now and were rather thrilled that we were going to visit him. But I dreaded facing Mike Handal, the Secretary of the Mission, who had seemed quite formal to me, dressed in his Sunday best Western clothes. However, I felt the Spirit sweetly upon me as he came out to meet us. The children were lined up rather solemnly—the boys in the yard and the girls inside. It is quite an important thing to them to have visitors, I guess. When they caught sight of me in the headdress, they all burst into giggles and gestures, indicating surprise and amusement. Mr. Handal gave them a stern look, but I turned and greeted them in Arabic and said my little piece about being a Bethlehem lady. And this really broke them up.

Inside Awni was all polished up and waiting. He bowed quite stiffly, looking frightened. But his younger sister came forward and greeted us in English, much more at ease. I said the lines I had memorized so carefully. They went something like this, "Hello, Awni, I have come to see you. And I have a gift for you." He took it with trembling hands. And Mike had to help him open it. After a few moments of rather stilted conversation, Mike called the girls' choir to come in. They were waiting in the hall, from which we could hear quite a few giggles and loud whispering. There were about twenty-five of them and I loved every one on sight. One of them kept giving me luscious doe-eyed smiles, and seemed especially eager for me to love her. But all were warm and most desirous of pleasing us with their songs. Without any accompaniment they sang together in good harmony and in English. It pains me that I cannot remember exactly what they sang. But I believe the first number was "Master, the Tempest is Raging." And there was a chorus of "Living for Jesus," sung much as we sing it here. After that Dwight asked them if they wanted me to sing, and they did. So I did! But what I sang has slipped my mind completely. I know it was about Jesus and that they seemed to like it. And I believe we then sang together "Jesus Loves Me." By this time I was truly swept out of myself with a flood of love for those darlings. I could feel how easily they could be led into the Baptism of the Holy Spirit and a fervent love for Jesus. I also felt their hunger for love. I learned later that their "Mother" at the Mission had left because of illness, and that they were short of supervision. Mike asked two of them to take me on a tour of the mission, and the loving-eyed one rushed to me, threw her arms around me and was promptly reproved for it by the Secretary. But I told her it was all right, and hugged her right back. Most of them trailed after us as we went through the schoolrooms and dormitories. They were very poorly furnished and not too clean appearing. But I recalled that in this poor Arab country they were likely much better than anything these refugee children had ever known. They were all clothed well and looked healthy. The very little ones, all in the care of Arab women, were most appealing. I laid hands on as many as I could, and blessed them, feeling led to do so.

There was not time enough for us to visit the Crippled Children's Home, the Vocational School or the Mission church. And it was so very hot that the thought of walking even a few blocks was overwhelming. Mr. Handal urged us to come again and stay for a real visit in Bethlehem. He said there was much in Jordan he would like to show us, and if we should come he would gladly guide us to all the holy places. He would expect nothing for this, he said, except that we tell others about the work of the Christian Approach Mission, since prayer and help is so sorely needed. I too was feeling by this time that prayer was truly needed. I know this Mission has done much good there. The King himself has recognized it, and his mother has visited it and commended them, as have other notables. Yet I felt a great lack in a *spiritual* way.

I went away with a heavy burden, which I ask you to share. Let us believe God for a real outpouring of the Holy Spirit upon the leaders and children there!

The Spirit spoke to me to adopt another child for us on this Mother's Day, since we have lost our blind Souad. I found that four needed sponsors. How hopeful they each looked! I wanted to take all four, but knew I must choose, so I asked the Lord to do so. And He selected a dear child who said her name was NOEL! How fitting—a Noel in Bethlehem! She spoke almost no English, but she took my hand, bent low, and kissed my wrist and fingers, giving me an Arab salute as she raised her head, by touching it with her hand. Mike said this is a sign they give only to those they love and respect. I embraced her and told her I loved her and that she would have about twenty-five more sponsors in America, trying to explain that we were adopting her together. Mike told her all this in Arabic and she beamed with joy. Then Dwight took our picture. She is a darling child. Before I left I also gave a special gift to the Mission so that they could take another child. They have a waiting list of over 1,000. It seemed fitting that the first offering I gave on our Tour was here where Jesus was born.

All too soon it was time to go! But before I left I felt to ask for a drink of water. I needed it badly. But I also recalled how David longed for water from Bethlehem's well. And in memory of him I wanted to drink a little of it. It was lukewarm and not too clean appearing, but it was holy water to me! Mike told us what great good had come from the purchase of the Water Truck. They had not only their own mission and school supplied, but also the local hospital and other missions. They had made many friends for Christ with that, he felt. And he thanked me again for the part we had played in helping to buy the truck. I recall that this gift was a portion of the sacrifice offering we had made at Epiphany in 1960, when we first decided not to give each other gifts at Christmas, but use the money for Jesus and the needs of His children instead. How wonderful to help supply water to Bethlehem! Mike went on to say that sufficient water is now flowing in Bethlehem and there is no longer a need to call it. But the truck stands by in case of emergency. "We would have had to close the mission," Mike said, "had not we been able to get the truck." I truly believe the Lord must have answered prayer about the water, for Bethlehem has been in desperate need of water for a long time. And we learned later that Jerusalem and other nearby places are suffering from lack of water, while Bethlehem has plenty.

It was with reluctance that we got into the taxi Mike had kindly ordered to take us back to Jerusalem. I longed to linger and visit all the Mission work, walk in the streets, learn to know and love the people—and really worship there. Yet, by the power of the Spirit, I know that I was able to love and worship and witness in various ways during that brief hour. And I learned later that the witness given among our own people on the bus had already been very effective. All praise to the wonderful Holy Spirit!

Yet in myself I felt frustrated and most empty as the taxi began whirling around the curves back toward Jerusalem. I longed for silence, in which to meditate and worship. But silence is a rare commodity indeed on a Tour. The taxi driver was most talkative. And from him we heard the same sort of complaints we had already heard repeatedly—the sufferings and losses brought upon them by Israel, and the terrible injustice of allowing them to have the land. I was glad I had really studied about the Israeli-Jordanic situation before visiting there. And I knew there were two sides to the picture. But we were kind toward this man—a refugee himself—and felt to manifest love toward him. I noticed a variety of pretty wildflowers on every side when he would occasionally slow down for a sharp curve. How they do drive in Jordan! All that has been written about it is true—and perhaps more. But some way they make it!

About halfway to Jerusalem we ran into a large flock of sheep being led along by a gay Arab shepherd. It was so picturesque a sight that Dwight could not resist a few pictures. So he asked the driver to stop. The shepherd knew the driver, so agreed to pose (of course Dwight gave him a tip.) When it was time for the flock to move on, I was amazed to hear a loud hiss or two—to which every sheep responded instantly. Then my heart filled with joy. I remembered in Zechariah where the Lord had said that He would "hiss" for Israel, and gather them. Most of the commentators have said that this referred to a special "whistle." And I had learned that only around Jerusalem did the ancient shepherds whistle for their sheep. And each shepherd's whistle was distinguishable to his flock. However in other places the shepherds called the sheep with their voices.

It was indeed precious to have heard a shepherd's whistle—as we had in the city. But here was one who really *hissed*—a sort of combination hiss and whistle! I was taken at once into thoughts of Zechariah. I remembered that I had been shown to honor him especially while in Jerusalem, and also to bring forth parts of his prophecy. I wanted to visit his tomb, but was not sure where it was. And I knew that it was not definite as to which Zechariah that it honored. I sank into a sort of reverie about him—and seemed to travel with him the last few miles back to The Holy City.

It was not until we turned into the Jerusalem streets that I came back to myself and realized that I was very hungry, thirsty, tired and weak. It was after two o'clock. So it had been over six hours since we had eaten. But, praise God, we had visited the Mission and seen Awni, in addition to all the thrilling places of interest. I lifted up my heart in praise to God for all He had handled so beautifully, and I told Him I was depending on Him to get us to the Mattars and the Mount of Olives Chapel soon also. Then I dragged my exhausted body out of the taxi and into the hotel. But my heart was still back in "The City of David" singing Christmas carols.

"CHRISTMAS EVE JOURNEY TO BETHLEHEM"

There are few people of my acquaintance who have not said when Christmas draws near, "How I would like to be in Bethlehem on Christmas Eve." Perhaps the experience would not be what you have anticipated, and perhaps your illusion that Bethlehem remains a tiny, secluded hilltop village, would be spoiled. The modern world is making innovations into this land of the Bible, but the land and its people retain much of their old world flavor and charm. A visit to the little city of David during the Christmas season, or any other time of the year, is an unforgettable experience.

Why not pretend as you read this article that it is the day before Christmas and you are coming with us on a journey along the winding macadam road from Jerusalem southward to the little city of Bethlehem. The road runs very close to No Man's Land, dividing Jordan from Israel and we watch the crooked strings of barbed wire with certain feelings that they have no business spoiling the beauty of the landscape. The countryside, although arid, has a beauty of its own and the horizon is a wide, unbroken expanse of hills meeting the blue sky. The contours of the mountains have been molded by the wind, and every hillside is crisscrossed

159

by the narrow tracks of countless flocks of sheep and goats. Driving along we are sure to pass flocks of sheep, see a caravan of camels in the distance and villagers with their children walking along the roadside. Seeing the little city of Bethlehem perched proudly on top of the long range of hills ahead, you know instinctively before the car begins its last winding climb, that you are approaching the town that gave us Christmas.

Before visiting the famous fourth century church, we decide to drive to the Shepherds' Fields for the annual Christmas eve service held by the Jerusalem Y.M.C.A. Skirting Bethlehem, our driver descends into the valley, really a plateau, follows the road through the village of Beit Sahur (village of the shepherds) and we stop before a gate and fenced-in grove of pine and cypress trees. These hills are honeycombed with caves, and in this secluded grove there is one special large cave which commemorates this whole area as the place "where shepherds watched their flocks by night." We walk through the whispering pines with people from many lands gathered to worship on the eve of Christ's birth. The outdoor service is simple, impressive and long to be remembered, conducted by the Y.M.C.A. and visiting ministers. The service is followed by a simple meal of lamb and bread prepared by shepherds from Beit Sahur.

We decide it will enhance our appreciation of Christmas in Bethlehem to make the return journey on foot from the Shepherds' Fields. There is time to spare as it is just twilight and services in the Church of the Nativity do not commence until eleven o'clock. The deep purple of night will descend quickly as it does in the Middle East, so we begin our walk. The fields of Boaz become a dark checkerboard in the fading light and soon there is a glow crowning the hilltop made by the myriad lights of the little town of Bethlehem. Even as the shepherds of old, we are making a pilgrimage to the city of David. Rounding the last curve of the steep hill which has passed the village well and the town of Beit Sahur, we mingle with thousands of other pilgrims. We enter the Church of the Nativity and go down the two flights of stone steps that lead into the Grotto of the Nativity. We are humbled in the hushed atmosphere of this sanctuary. This may not be the exact spot, we realize. Yet, we hold it in reverence because this place has been venerated by Christians of every generation since 330 A.D. Then we make our way to the Latin Church where the traditional Christmas Eve service is held. We ourselves may worship in a different way, but on this night it seems fitting that we should join hands and hearts with people of other faiths who love the Lord Jesus as we do, yet worship Him in a different manner. At midnight the bells of the churches of Bethlehem ring their message of peace and goodwill to all the world. Midnight is passed. Christmas Day is here. Hallelujah, Christ is born!

—Portions of an article from *The Palestine News.*

THE VIA DOLOROSA

A t three o'clock we are all going up The Via Dolorosa, so hurry and eat your dinner!" This was the greeting we received from one of our Tour ladies, as we entered the hotel. Glancing around, I could see a number sitting in the lobby writing cards or reading. I was pierced by the realization that Dwight and I had kept them all waiting. But I learned later that they appreciated time out for a little rest and relaxation. Meanwhile there was no rest-time for us. Hurry! Hurry! Hurry! My body and mind were beginning to rebel at being constantly rushed and pushed about. But the Lord gave grace—and besides we were very hungry and didn't mind cleaning up in haste and rushing to the dining room. How delicious everything tasted! I forgot that I was eating only half portions and ate freely of everything in sight. And the waiters for once were very prompt. All the other Tour members had eaten, so we enjoyed just being alone together. We agreed that so far the meals had been much better than the travel book had led us to expect. But we wondered about the strange pungent odor and taste that seem to permeate the dining room and hotel. It was not like any other in my experience. "Do you suppose it is some kind of spice?" I commented. And I recalled that as a child I had stayed in a mountain resort where the cook flavored everything, even the breakfast eggs, with bayleaf. But this was certainly not bayleaf. Even my bitter tonic now tasted of this Jordanian flavor! So we had a good laugh about it and decided it must be all right, though it was a little monotonous to have everything taste alike.

We made it by three, or shortly thereafter! And this time I had on my sunhat. It was still very hot at that hour, and the city was dazzling to our eyes, as the sunlight reflected on the gray-white stones. I wish I could remember clearly all that took place that afternoon, but it is a blurred jumble in my mind, perhaps due to my fatigue and the disrupting confusion in our group.

Our guide, Ahman, was patiently and even reverently trying to tell us about the Via Dolorosa, as we walked through the Sheep Gate again and on inside the walled city. But our people, particularly some of the Ministerial brethren, were plainly not interested. They kept making disparaging remarks about things being unscriptural and Catholic. And our guide was confused and pained. He was not used to Christians being disrespectful about The Way of the Cross. I was glad I had read about the Via Dolorosa and had seen it pictured several times. I knew, of course, that the actual streets on which Jesus walked were buried some thirty to sixty feet under the ancient stones on which we would walk. I realized too that the fourteen Stations of the Cross, so dear to the Catholics and Orthodox Christians, were partially scriptural and partially traditional. But I had hoped that we could make this pilgrimage with the right attitude—even though this would be difficult to do on a baking hot day in late May. Oh, to have been there at Passover time!

In my hand I was carrying a rosary our friend had asked me to take with me for him as I climbed the Via Dolorosa. I realized how very much he desired to make this pilgrimage himself and how reverently he would have approached it! Therefore, as I carried his rosary in my hand and him, as well as you, my dear sisters, in my heart, I wanted to be a worthy pilgrim. But, try as I would, I could not seem to get my mind on Jesus and the Cross. For one thing, the people were thronging around us—so crowded were the narrow streets. And the sickening odors of the walled city—where thousands of refugees lived in "cubbyholes" in the old stone buildings, without any modern sanitation or running water—were heavy around us. I tried to praise the Lord and have grace, but I was constantly nauseated. Yet this sickness was mild in comparison to the heart-pain I felt for these people, particularly the children. Oh, how I longed some way to turn them all to Christ and to minister to their obvious needs!

Then, suddenly, I realized that the city streets through which Jesus passed must have been much like these—crowded, dirty and smelly. And His heart-pain for the people was greater than the physical suffering He bore on that dreadful day. So, perhaps after all, I was in fellowship with Him as I bleakly followed our guide to the beginning of The Via Dolorosa—The Church of the Flagellation. By this time most of our crowd had straggled off and were either taking pictures or going into the shops. Dwight was nowhere to be seen—and I was tempted to get upset at him, upset at our group, and just upset in general. Our poor guide didn't quite know what to do, but tried to continue on with the few who would listen. He pointed out to us the remains of one of the two Eastern towers of the famous Fortress Antonia, which was the headquarters of the Roman Army in Jerusalem in Jesus' time. On the former site of this fortress three buildings now stand—the Flagellation Convent, the Al' Omariyeh College (Moslem) and Sion Convent.

I had read about Sion Convent in Dr. Field's book, and in other accounts, and was thrilled at the discovery of the actual Lithostrotos—where the soldiers mocked and lashed Jesus—beneath the building. I knew that the sisters there were always glad to show visitors the ancient stone pavement, on which were carved the games the soldiers played. The marks are still visible! But our guide led us on toward the college, and spent considerable time there telling us how wonderful it was and stressing many Moslem factors. The Praetorium, from which Jesus was led forth toward Calvary, was located near or on this ground. Although Jesus received His death sentence on the Lithostrotos, it is the custom to begin the Way of the Cross at this college. The second station, where Jesus received the Cross, is fixed on the road outside, opposite the Chapel of the Condemnation. Several of our people had informed Ahman that we didn't want to see any more Catholic Churches, so he hurried us along to the Third Station—where Jesus fell for the first time. It is under the so-called Ecce Homo Arch. This is a Roman Arch, traditionally associated with the Way of the Cross and with Pilate's words: "Behold the man!" It is the middle arch of a gate, said to have been built by Hadrian. The Fourth Station, where Jesus caught sight of His heartbroken mother, is marked on the street by a bust of the Virgin and Child, placed in front of a small oratory. Nearby are an Armenian Catholic Church and a Polish chapel and museum.

Our guide told us that every Friday afternoon at 3.00 p.m. the Franciscans lead a procession along the Via Dolorosa, and pilgrims and visitors join in. It was difficult for me to picture any sort of procession making its way through these narrow crowded streets. But perhaps the people move out of the way for it. We kept running into children, sheep, donkeys and Arabs, besides all the tourists. At the Fifth Station, it is said that Simon of Cyrene was compelled to carry Jesus' cross. This station is marked by a Franciscan monastery. At the Sixth Station, Veronica is said to have wiped the face of Jesus, and to have received the imprint of His face on her cloth. A Greek Catholic chapel stands here. Still climbing steadily uphill, we came to the Seventh Station, where Jesus fell for the second time. The Franciscans have two chapels here. As you can readily see, the Via is lined with places of worship, yet sandwiched in between are many old houses, each one sheltering multiple families in very crowded conditions. We had already crossed King Solomon Street and soon found ourselves on St. Francis Street. (There is also a David St., a Christian St., and a St. of the Chain that I noticed.) The Eighth Station is marked by a cross on the wall of a Greek Orthodox Convent. At this place Jesus spoke to the weeping daughters of Jerusalem. At the Ninth Station Jesus is said to have fallen for the third time. A column marks the place. For some reason a detour had to be made along here and I got confused and lost the guide. Still clutching the rosary, but being very careful that none of our good Pentecostal people saw it, I stopped to get my bearings. I knew that the last five stations were in the Church of the Holy Sepulchre. Try as I would, I could not feel that the hill of Calvary was inside that church. Nor could I believe that Jesus' tomb was there—as so many millions do. Nevertheless I realized that it is a very ancient Church, highly revered and loved by multitudes of Christians. So I wanted to enter it in the right spirit. I tried to praise and draw near the Lord. But at that moment a dear colored sister attached herself to me. (I had run into difficulties getting away from various chattering or complaining sisters during this trek, but I sensed that this sister was in the Spirit.) Her heart was broken, she said, at the carnal attitudes and ways of our Group. She spoke of having heard me singing in the Spirit as we journeyed into the city, and said how she wished everyone had really worshiped also. She felt the Lord was very grieved with us. There, in the midst of the Old City, in the stifling heat of the afternoon we stood, and as she lifted up her voice she reminded me of the ancient prophets who lamented about Israel's hardness of heart in much the same manner. Human nature doesn't change! She spoke of the coming Convention at Pentecost, and dolefully predicted that if we didn't repent and change before then, the Lord surely couldn't pour out His Spirit upon us. Then she started to weep and said, "And if He doesn't meet us there, we're sunk—we've had it!" "Oh no," I cried, "The Lord is going to pour out His Spirit in a greater way, whether at the Convention or in other places and times. But He will choose vessels that are prepared and expectant." I still remember how clearly I felt the truth of these words, and I too felt like a prophet as we turned to walk slowly into the entrance of the Church of the Holy Sepulchre. Many deem this the most sacred of all the Holy Places in the Holy Land.

Inside the Guide was already giving his lecture about it, so I had to read later in the guidebook what I had missed hearing. In A.D. 135 The Emperor Hadrian made this area into a forum and marketplace, placing pagan statues all around it. But in A.D. 326, when St. Helena arrived in Jerusalem, the Bishop of Jerusalem told her that beneath this forum Calvary and the tomb of Christ were located. Constantine ordered the area cleared, and they found caves and an apparent tomb cut into the rock. Some crosses also were found and one was identified as the Lord's. Around this tomb the original church was constructed, and apparently everyone accepted this site as genuine, even though it lies *within* the city walls, whereas the Bible states that Jesus was crucified *outside* the city walls. And it is known that no burials were allowed within the city. However an explanation is offered for this—that the walls have been moved farther out. Yet remains of the ancient walls have been discovered. And they plainly show that the site of the church is within, not without, the walls! In addition to the church, a basilica and vast court were also constructed, the latter built around the rock which they said marked Mount Calvary. Beneath the forum they supposedly found the Cross of Jesus and the two thieves.

In 614 the Persians destroyed these three edifices, but they were partially restored in 629. Again in 1009 the buildings were destroyed, this time by the Egyptians. In 1048 The Emperor Constantine Monomachus rebuilt the church. And the Crusaders further added to and improved the building, making several sanctuaries all under the one roof. In 1808 the church was destroyed by fire. A Greek Architect very poorly carried on the work of repairing it and as a result "there are shapeless additions and senseless mutilations." In 1869 the great dome was replaced. Another fire damaged it in 1949. Both earthquakes and war have also damaged it. It seems to be an ill-fated building indeed! At present it is being supported without and within by huge scaffolds and braces. And it took a little courage for me to go inside, knowing how severe the earthquakes can be in Jerusalem. I hoped one wasn't about due!

The only entrance is in charge of two Jerusalem Moslem families. It is said that although the key could have been given over to the Christians, when the dominion of the Turks was ended, they preferred it to be in the keeping of the neutral Moslems, who deem it a great honor to hold it—and strive to keep peace between the Christians. The following Religious Groups have altars or chapels within: Roman Catholic, Greek Orthodox, Armenian Orthodox, Copt, Syrian Orthodox and Abyssinian. In March 1958 major repairs began on the Church, and these are now in progress.

The first thing we saw after entering were the Moslem keepers. Then the Stone of Anointing was pointed out—where Jesus is said to have been laid when He was anointed for burial. I have a dim memory of following the guide up some stairs and to a little chapel. By this time several of our noisy brethren had joined us, and general confusion was the result. They were critical and skeptical of everything the guide said, and asked him to back up his statements with the Scripture. This so muddled him that we seemed only to mill around from one altar to another. Try as I did, I did not feel very reverent. However, when we came to the Greek Orthodox altar, a kindly priest stepped forward and offered us lighted candles. Several of our number rudely refused them. But a few of the sisters took one, and I felt led to do so too—even though the brothers began to criticize us and say we were "going" Catholic. Yet I have to admit that as much as I have always liked candles, and felt them in connection with worship, I was getting a little sick of them by now—for they were in evidence at every turn. And it was very difficult to see in so dim a light. Still clutching the rosary in one hand and the candle in another, I started down some very steep winding stairs. A strange odor aroused me from a sort of daze, and I realized that I, or rather my hat, was on fire! About this time some of the brothers saw it too, and began yelling most irreverently. One said, "It serves you right for taking a heathen candle." I was really aghast—not at the fire, but at their rudeness. For of course I quickly snatched off my hat and put out the fire at once. I felt all shaken up, for I knew our clamoring voices had been heard by many within the Sanctuary, and no doubt they wondered what sort of crude people we were. Secretly, though, I think I was a little thrilled by this fire, for in this Church the famous ceremony of The Holy Fire is held each Easter! (The Greek Patriarch goes into the tomb, while the multitudes without and within worship and sing. Supposedly his candle is lighted by fire from heaven. He then comes out and lights other candles, and it spreads until everyone's candle is aflame.) However, Dwight had another explanation—he said that perhaps I had carried some Pentecostal fire along with me.

The designated tomb of Christ is approached by a little marble Chapel of the Angel, marking the place where the Angel is supposed to have appeared to the women. Only two or three may enter the Tomb at one time. So we had to line up to take our turns. And of course one could hardly linger, with so many waiting to enter. There were other chapels too, which I vaguely remember—one is called The Chapel of Adam, for his skull is supposedly buried beneath it—hence Golgotha—"The Place of the Skull." It is believed that the blood of Jesus fell into the earth and reached Adam's skull. There is also an altar dedicated to Melchizedek, the first King-Priest of Jerusalem. There is a stone chalice placed in The Catholicon—supposedly marking the exact center of the earth. Among the many other chapels is one for Mary Magdalene, to whom Jesus appeared, and one for Mary, His mother.

I couldn't quite place the last five stations of the Cross—but they are all within this massive building. In my confusion, I suddenly recalled how the blessed Holy Spirit had taken me in a very real way along the Via Dolorosa a few years ago. I had lingered at each station for several days. He had given me a song or poem about each—and had also revealed not only the historic meaning of the stations, but had also given me present-day truth and inspiration. How very real that pilgrimage was! And how superior to the actual pilgrimage I made on this hot day in May!

On the whole my impression of the Church of the Holy Sepulchre was depressing and confused. I was grateful for the delightful coolness of it, after the heat of the streets. But I felt no sense of blessing or exaltation in our visit there. After recalling my own impressions, I found similar sentiments recorded by Bishop Fulton Sheen in his fine book, *This is the Holy Land*. He says of it: "This was one of the most beautiful and unusual churches the world has seen; today it is the most painful spectacle in the Holy land . . . Its history has been one of steady disintegration, culminating in 1927 in an earthquake which threatened to bring down the entire building . . . Now, thirty-three years after the earthquake—the venerable structure is still theoretically held together by supports which actually, in contraction and expansion during the heat of the Palestinian summer, are pulling apart what they were erected to protect. (I remembered reading that engineers say it is held up only by the grace of God.) One's first impression of the Church is sadly different from that of the fortunate ones who saw it in all the noble clarity of its original design.

Now one sees a dark warren where gloomy flights of steps lead here and there into the heart of a pious labyrinth." According to him, the Franciscans have desired to restore it to the original design. But the other Church organizations, who have rights there, have voted only to try to preserve it as it is today.

When we finally emerged from its gloom out into the still brilliantly lighted streets, we all seemed to feel let down and very weary. Our guides were acting as though this was the end of the tour. But some of our brethren began clamoring to be taken to "The Garden of the Tomb." "Now THAT'S the place we want to see," they kept chorusing. And for once I was glad they were persistent. At first Ahman hesitated. I believe these Moslem guides tend to stick closely to Catholic tradition. And that Garden Tomb is Protestant in its administration. At the hope of going there, I revived instantly. "It's quite a walk," our guide told us, while we groaned. But we made it clear that walk or no walk, we wanted to go there. While he counted noses and hunted for a few lost sheep, I found time to dash into a little bazaar booth that had that wonderful American commodity—bobby pins. I bought a package with joy, hoping now to do something with my hair, which kept growing more and more unkempt, if that were possible. Unfortunately I didn't notice that there were only about a dozen in the package, and I needed three times that many!

The peculiar odor, of which I spoke, was heavy around us at every turn. Coming out of this little store, I entered into conversation with another tourist, and she informed me that the odor is that of sheep fat. They use it for all their cooking, she said, and for other purposes too. It occurred to me that it must be somewhat similar to the savor of the Temple, in the days when sheep were being constantly offered unto the Lord. I began to enjoy it!

While Ahman searched for our lost sheep—I took a good look at the Citadel, which I had hoped to visit. Even at a distance its three towers are easily distinguishable. Each time I had seen a picture of the Tower of David (another name for the Citadel) I had felt a sense of identity with it. This Citadel occupies in part the site of the palace of Herod the Great. It stands close to the now closed Jaffa Gate. David Street begins here. Herod built three massive towers there—and Jesus must often have seen them. Titus did not destroy these. But Hadrian had them dismantled partially and rebuilt. Saladin later reconstructed it. When General Allenby read the King's proclamation, freeing Palestine from the Turks in 1917, he stood upon the eastern steps of the Tower of David. I wanted to stand there too and praise God and declare the Word of a greater King!

Achiral! (At last!) we were all assembled, and Ahman led us through the majestic Damascus Gate—the most impressive of them all. We straggled after him like tired sheep. I was trying to remember that many blessed saints, dignitaries and famous people had used this gate. But this was too much for my sense of humor, which suddenly revived, and I exited laughing, thinking how ridiculous unsaintly, undignified and even funny *we* all looked! Some of the men were wearing Arab headdress, and so were a few of the women. Others had on bright scarves they had bought from street peddlers. I was still wearing my singed sunhat! Many a strange pilgrim band has made its way up to that Garden of the Tomb. And I realized that day that we Americans too are strangely lacking in grace and manners—too rich, too proud and too high-minded to be well loved abroad. Then my laughter quickly turned to tears, for it was all too evident that our guides and the people around us were not seeing Christ in us at all, but just some spoiled Americans.

Each step of that uphill climb, which proved not very long after all, brought us nearer to the beloved Garden of the Tomb. And it was amazing the way our spirits arose, in spite of the heat and our thirst. When we turned into a walled alleyway and saw the gate ahead, several broke forth into exclamations. And again I felt to lift up my voice in praise and thanksgiving. In a moment now we would pass through the Gate. And a dear Arab waiting for us there must be Brother Mattar—whom we had been trying so hard to meet. Perhaps we would also meet Jesus—in a precious way—in the Garden! With this high hope we hastened on toward the entrance.

THE GARDEN TOMB

My first impression about The Garden Tomb dates back some seventeen or eighteen years when, one day, I was rapidly leafing through a copy of *The Defender*, and my eye was caught by picture of it. I slowed down, gazed intently at it for a few moments, and then began to read the article. A sweet warm glow infused my heart, and a quickening ran through my being! I felt certain that I was reading about the actual site of the Crucifixion, burial and resurrection of our Lord. I realized then, with a shock, that I knew nothing factual about Jerusalem and the Holy Land. All my concepts were based on Sunday School pictures, movies about Bible stories, or my own imagination. It seems incredible now that there was ever a time in my life when I did not really know the Holy City and Land. It is certainly true that it is the Fifth Gospel, and must be "read" carefully and lovingly, if one is to understand the other four. But my knowledge of the Land is only recent. In fact, I had no idea, as I first read about the Garden Tomb, that there was another sepulchre in Jerusalem that was accepted by all the Orthodox Church world as Christ's burial place, and that this was located in The Church of the Holy Sepulchre. So you can see the extent of my ignorance!

I am sure this first impression must have been very deep, for I can readily recall it at will. It was followed by other concepts, mostly based on additional articles published in *The Defender*. Then, when, about five years ago, Dr. McKoy began to write about it also and to tell of his visits there, and his participation in Easter Services—I began to feel that the Garden of the Tomb was very much a part of us, and I truly longed to visit it. I was glad to find so fine a description of it in Dr. Field's book also, and in other books I read before starting on our journey. Several writers had said that it was the dearest spot in all Jerusalem, the place where one could feel the presence of Christ in a most unusual way. Of course these writers were Protestants, for no Catholic would dare to disagree with the edict of the Church that claims authenticity for the other site, which I have already described in a previous writing.

But I found that many fine scholars are among those who believe that The Garden Tomb is authentic. And a number of books have been written to prove this theory.

As we approached the entrance to the Garden late in the afternoon last May, I had a feeling which can be likened to the mood of the song, "The Last Mile of the Way." We had been to the Mount of Olives, to the Garden of Gethsemane, to Bethlehem, and through the Old City, following the Via Dolorosa. We had visited the Church of the Holy Sepulchre, with its vast confusion of labyrinths and altars, memorials and all. Then we had followed our guides on out through the Damascus Gate and up the road to this Garden.

> "When I've gone the last mile of the way,
> I shall rest at the close of the day,
> And I know there are joys that await me,
> When I've gone the last mile of the way."

The song speaks also of seeing "the Great King in His beauty." I am sure my heart was skipping as much from anticipation as from exertion, when I slipped into the Garden at last, and stood alone for a moment in worship, letting my eyes feast upon the open tomb. Then there were greetings, most joyful, with the dear Mattars. And when they learned who we were, their Arab faces glowed with welcome. Brother Mattar asked us to remain after the others left, so we might really become acquainted in person. (Of course we had been writing to them for some time, and had sent contributions to the new Chapel on the Mount of Olives.) Our arrival in the Garden had been timed by the Lord, I am sure. They had not expected us, but were free. One group had just left and there was about an hour's lull before the next was due. It had been a full day in the Garden. Ever since sunrise visitors had been coming and going. Some came alone, hoping for quiet meditation. Others came in small groups—or in larger ones. Most of the latter wanted to hold a service there. So Brother Mattar had told the crucifixion and resurrection story again and again, always with the open Bible in his hand, though he spoke from memory. Yet he was apparently delighted to hold a service with us, and to tell it once more. If he was hot and weary, his face showed no sign of it, and his voice rang out with joy and vibrant faith, as he began to repeat the beloved, familiar words to us. (I learned only later, from his wife, how very long his hours are and how completely he expends himself in this testimony and ministry.)

I am sure he found in us a most responsive and helpful congregation. When we had learned that the Garden would be free for only an hour, we had all found seats quickly in the little outdoor amphitheater that faces the tomb area, and is elevated above it, among the trees and flowers. A miraculous change had taken place in us! I hardly recognize many of our tour members. That Garden of the Tomb had become a real Garden of the Resurrection already! We were no longer tired, thirsty, disillusioned and heavy—but radiant with joy. As we had settled into our places, we had spontaneously praised God and sung choruses to the Lord. The atmosphere of the real Pentecostal service settled around us. But there was something more in the air! It was intangible, but none the less real. I am sure it was a real visitation from our Lord, a most unusual moving of the Spirit that all of us seemed to feel and embrace. Weeks later, as our Tour neared its end, I recall that several rehearsed again our meeting in the Garden Tomb, and they said that it had been the most glorious service they had experienced during the trip, even including those at the Convention. In fact, they agreed, it was the high spot of the journey!

Whether or not this remarkable presence of the Lord was an indication that this Tomb and Garden are indeed HIS, and that His presence and blessing still linger there, I cannot say. It is possible that He met us so openly because we were weary, worn and hungry for Him. Yet, I am sure it was the general conclusion of the company that this *witness of the Spirit was indicative of the authenticity of the Tomb.* I am glad that to us it does not really matter. If it is not the actual sepulchre, it most certainly is similar to it, for it answers the Bible description of it perfectly, and is the *only one in or near Jerusalem that does.* Sitting there, in the shade of the trees, smelling the faint fragrance of hollyhock, the wild anemone, stocks and other flowers similar, if not identical, to those grown in Southern California, I felt completely at home. And I pictured each of you dear ones sitting there with me. Yet I had a secret satisfaction in remembering that our Lord has made a rocky tomb on our mountain very much HIS, on several occasions, and that He had given us a Garden of the Tomb here, in which He met us even more gloriously and openly than

He was manifested that day in Jerusalem. (Who can ever forget that amazing Easter at the tomb, when everything turned Kingdom-gold in the sunrise, and we were transported to heavenly heights in Him?) We are sure that Jesus is not confined to places—though they are sacred and precious, when identified with Him. It is only the Holy Spirit who can take us to the tomb—and out of it in resurrection power! How wonderful to know Christ in the Fellowship of His sufferings, planted with Him in His resurrection, and risen with Him in His eternal life! We have been many times in the Garden of the Tomb, within our own hearts. And I found some of these memories darting through my mind as I sat in worship, listening to our brother bring forth the Word that day.

Our little song about the women hastening through the garden, lived before my eyes. It fitted perfectly!

> "Their hearts were sad and torn,
> Though their feet were swift and strong,
> As they hastened through the garden
> To Jesus, their love, on that bright and cloudless
>
> morn.
>
> Then, as they drew near,
> They were gripped by fear,
> The tomb was empty,—
> The stone rolled away!"

Suddenly, a surge of the Spirit swept through us! Praise arose like a fountain! I was almost overcome with joy, hearing the volume of praise. I sensed the Lord's desire to do something unusual, and I felt surely He would speak openly. Then pain pierced my heart. I knew instantly that Brother Mattar was NOT Pentecostal, as I had supposed. And I saw that he was uneasy and negative, as preachers become who fear being interrupted by the Spirit. He was looking at his watch. Suddenly he loudly interrupted the praise, saying we must go on to Golgotha, in order to make room for the next group in plenty of time. A sense of bondage and heaviness immediately descended upon us. You all know the feeling! But we dutifully quieted down and arose to follow him along the pathway to the "Place of the Skull."

There is no hill like this on earth anywhere! It resembles a skull most truly, and well merits its name. Pictures taken a long time ago reveal that the visage has been somewhat marred by frequent earthquakes. Yet, praise God, it has been preserved sufficiently for identification. The area in front of this skull is utterly barren and rocky. It is believed that this was the real place of stoning—where prophets and saints, as well as evildoers, suffered death by stoning, according to Jewish law. In Roman times it became a place of public execution, and there is every reason to believe that upon *the cranium of the skull* Jesus was crucified, for this is the literal meaning of Golgotha. (The word Calvary occurs in only one place in our Bible and there it is a mistranslation, according to scholars.) There is another thrilling fact concerning this hill that leaves one almost breathless! Ancient maps, as well as underground excavation (in digging limestone), reveal that this hill is the northern peak of Mt. Moriah. (The sacrifices in the Temple were said to have been made on the northern side, and other Bible records show that the Father chose this mountain for the Sacrifice of His Son. In offering Isaac, Abraham was directed to Mt. Moriah— meaning YHVH-Jireh, "the Lord will see and provide," or "on this mountain the Lord will be seen.") David also was directed to buy this area for an altar. And God's judgment was halted here. (1 Chron. 21:15) This mountain's northwest corner was cut away from the eastern portion when a large moat was dug just outside the wall. Recent excavations have revealed that the wall did extend to the Damascus Gate, at the time of Christ, and that He most likely passed through this very gate on the way to crucifixion. Golgotha should lie just outside the walls, and this Mount of the Skull does! Furthermore it shows evidence of the most severe earthquake in which all its huge rocks were actually rent in two, as the Bible says. We all remember that the Old Testament sin offering, after being slain on the north side of the altar, was taken outside the north wall of the city and burned. Christ was the antitype of this sacrifice, so it was necessary for Him to die outside the city, yet on Mt. Moriah, to the north of the Temple.

As we stood gazing in awe at the Skull, Brother Mattar related some of these facts to us. But I was too involved in the Spirit to hear him. So I later studied about them in the books he gave me. Jeremiah's cave is close by, too, to the right. The Garden Tomb is nearby—to the left, as we faced Golgotha that day. However, none of us could climb up on the cranium of that, where the sacred blood of the Lamb likely was spilt. This is because there is a Moslem cemetery on top, and no unbeliever may enter it. (There are a few exceptions of course, based on authority or money.) This provides a marvelous protection for the area, for most likely ardent pilgrims would carry off everything in sight, as they used to do. However, we were welcome to pick up stones, seeds or other tokens in the Garden, and so, of course, I did!

It was only a few steps back to the huge underground church which has been recently discovered in this area. At first it was thought that this was an ancient cistern, which archaeologists estimate was dug in about 600 B.C. The Jews may have used it as a refuge in times of persecution. But it has many Christian marks in it also, including a large cross carved on the walls, which mysteriously glows in the dark. Dwight climbed down the ladder about sixty feet and saw this himself. There is a little water in the bottom now, so he could not walk around it. But he verified the report that it is indeed huge in size. Christian artifacts have been found in it, and other evidences of its being used as a secret place of worship. If permission and financing are secured, further excavation will be done in this area.

The Tomb itself was hidden under piles of rubbish until it was accidentally discovered less than 100 years ago—in 1867. Excavations revealed that about A.D. 500 a Christian Church was built in front of the Tomb, the builders utilizing the rock face as the north wall. It appears that this Church may have been built in reverence for the Tomb, since the entrance was immediately opposite to it, and it would be seen first on entering. There was a heart-shaped baptismal pool, in which the convert could look directly into the door of the tomb as he went down into the water. The original entrance to the Tomb was very small, but has been increased now to a height of 5'9". One of the most remarkable evidences is a very deeply worn foot trail, made by the passing in and out of countless people. Since tombs were seldom entered, this would indicate that this particular tomb was venerated as a shrine.

The tomb is typically Jewish, hewn out of the rock, and that of a wealthy man, "He made His cradle with the poor in His birth, and His grave with the rich in His death." The interior is accurately worked to the cardinal points of the compass.

High up on the rock face at its entry there are four recesses or shelves. In removing debris around the entrance, objects of vile Venus worship were found. It is thought that these were once installed in these recesses. It is a recorded fact that Hadrian violated every place sacred to the Christians, and that he also violated Christ's tomb. Below the recesses on the left of the entrance is an incised symbol used by Christians at a very early date, probably about A.D. 200. It is the Christian Anchor sign, used often in connection with the Scripture, "Which hope we have as an anchor of the soul." It is evident that Christians worshipped here at that period.

The rock face of the Tomb was broken away long ago and the space is now filled in with masonry. Above this there is a window, and this bears perhaps the greatest witness of all. Very few ancient tombs had windows, and *no other known tomb has one made in such a manner as this.* Instead of it being placed in the side wall, this comes through the roof, slantwise, throwing a beam of light right across the Tomb, and reaching the furthest wall near the ground level within one of the three tomb-beds. Usually a tomb is quite dark, but this one is lightened from dawn till dusk, and *all day long a brilliant ray of light rests upon that bed*—the only finished bed in the Tomb. (The others were never completely hewn out.) The Bible story relates that when the disciples arrived at our Lord's grave early in the morning, they stooped down and looking in, they SAW! They saw the place where Jesus' body had lain; they saw it was not there, they saw grave clothes and the napkin. When the women came, they too saw! And Angels were sitting at the head and the foot of the grave where Jesus had lain. Two ledges are in this Tomb, located in just such a way as the Bible describes. The angels could have been seated thus on them. There is a small chamber for mourning located just inside the entrance, and the finished tomb-bed is the only one that would have been visible from the door.

I longed to enter the Tomb alone, but we were told to go in a few at a time. Here again the picture taking went on and I felt nearer to tears than to praise, though I tried to feel reverent. I was surprised that some could joke even in the Tomb! But others were truly moved, and a number entered weeping. I walked for a short time about the Garden alone, after I came out. And I saw that the Mattars were very busy indeed selling souvenirs and books and photo slides. So I enjoyed a brief time alone with Jesus, and He spoke to my heart in a personal way. Then, as I circled around, I was drawn back to the Tomb. And from it I could hear fervent male voices rising, seemingly speaking in tongues. I wondered just which of our brothers had returned and were worshiping. I waited for them to leave before entering. And when two men finally emerged, I was indeed surprised, for they were not from our group at all! And were still talking in "tongues" which proved to be Swedish—their native language!

Now it was my opportunity to enter alone. Oh wonderful moment! I feel I came very near to the heart of the Garden Tomb in the priceless minutes that followed. I had a chance to kneel and worship, to sit on one of the ledges, where an angel may have sat, and to meditate quietly, picturing Jesus on that rocky bed. I am sure Resurrection Day will always have a richer meaning to me because of this "tryst in the Tomb." I left a token or two there, tightly wedged in a hidden part of the rocky interior. And when I finally had to withdraw, it was with regret. A new party were beginning to enter the Garden. Dwight, meanwhile, was conversing with Brother Mattar. And he had been generous in giving us pictures and books about the Garden. It was time to go, so he and his wife led us lovingly to the gate, after we had arranged to meet right after dinner. I expressed to them my joy in visiting the Garden Tomb. And I was glad to hear from his own mouth that the Association of English Christians who obtained the property in 1894 have provided in the Trust Deed, "That the Garden and Tomb be kept sacred as a quiet spot, and preserved on the one hand from desecration, and on the other hand from superstitious uses." I was led to speak to them then of my eagerness to visit the Mount of Olives Chapel also. They seemed a little dubious about taking us there by night, but when they found that we were going to leave the next day, they agreed to do so. I knew I simply must visit the chapel and that it too was a most significant appointment with the Lord, for He had given me a special song about it and had spoken clearly concerning it.

Now we were on our own again! The buses had picked up the other members of our party a half hour earlier. We had no idea where we were or how to get to the hotel, even though the Mattars tried to direct us. But we declined to call a cab. We both had a longing to walk again through the blessed streets and feel the pulse-beat of Jerusalem. Coming out again into the noise, heat and smells of the city, I could appreciate all the more the sequestered quiet, beauty and fragrance of the Garden of the Tomb. It truly is a spiritual and natural oasis. I could understand why General

Gordon and many other Christians had delighted to retire to it often, finding in it an atmosphere entirely unlike that of the city itself. The late shadows were falling across the walls when we past the exterior of the Damascus Gate, for the sun had set. We were not sure just which way to turn, so we sought for someone who looked like they might speak English.

As I turned around, I found myself looking directly into the tranquil and rather beautiful eyes of a young priest. It was evident that he and his even younger companion, also a priest, were watching us with unusual interest and waiting for a chance to speak to us. They eagerly began to converse, and in an instant of time both Dwight and I felt a rapport with them which we had not found with our Tour Companions. It was one of those high and rare moments of spiritual communion. They quickly changed their course and offered to walk with us to the hotel, saying that they knew of a shortcut. The Holy Spirit came upon me in such strength that I was indeed as one drunk with new wine. Love, praise and anointed words flowed out of me in a stream. Dwight was walking and conversing with the younger priest, who came from Canada, and was studying briefly in Jerusalem, and learned that he was soon to be placed in Nazareth. The priest with me was from Ireland, and had flaming hair as well as a flaming heart. How graciously and fervently he spoke of our Lord! I soon found myself telling him about our coming Pentecostal Convention, and then about the baptism itself, and the wonderful way the Holy Ghost has moved in us. He seemed spellbound, and would stop every few feet, asking questions, prodding me to say more. But I couldn't have stopped anyway!

I also told him of our love for Catholics, and of the understanding of them the Spirit has given to us. I showed him the rosary I had carried all the afternoon— through the Garden Tomb as well. I told him how the Holy Ghost is being poured out in new ways upon the Catholics. And I mentioned Father Pio too. He was almost incredulous with amazement one moment, and then, the next, he seemed to be joyfully receiving my testimony. All too soon we approached the hotel! I felt led to have us stop short of it for our farewells and, as we did so, Dwight suddenly asked the priest with me to pray and to bless some little tokens we had obtained for two Catholic friends. Then, in a most Pentecostal way, the four of us linked our hands and stood in the middle of the street, lifting up our faces and voices to the Lord in love and praise. We seemed to be completely alone and free—and it was glorious! Before long, though, I was aware that some dozen Arabs had gathered around us and were watching us with both curiosity and wonder. So we turned to include them too in this effulgence of love, and when we smiled they too smiled eagerly and greeted us in Arabic. When I responded in their language, they were as pleased as children. We blessed them in Jesus' Name, and they seemed touched and grateful. Our dear priests were loathe to part with us. And the one who had walked with me told us that he would perhaps soon be sent to Los Angeles, and he took our card, just in case.

As we finally separated, both Dwight and I still felt intoxicated. "What would our dear Pentecostal preacher-brothers have thought of us," Dwight asked, "if they had seen us in such unity with two Catholic priests?" Of course we both knew the answer! And we still marvel that we felt actually closer to them than to those preachers! But of course, this was a special day in May, and we were in a special city, and Jesus met us in a special way! What a glorious ending to a spectacular day! And even now, as I write about it many months later, I feel the surge of that great love and joy and am sure that He, our Beloved, was truly in our midst in Resurrection Life, walking with us those few entrancing blocks through Old Jerusalem. And we, like two of old, knew that our hearts were burning within us, and wished that He would have tarried through the night.

THE MOUNT OF OLIVES CHAPEL

S ometimes a simple everyday act, like opening a gate or passing over the threshold of a door, assumes a surprisingly dramatic importance: time seems to stand still, or to leap forward or backward; or, perhaps, space loses its familiar identity, and we find ourselves in a whole new sphere! Before such a gate or door we are likely to pause intuitively, either breathless with eagerness to enter, or sadly reluctant to leave behind an experience which we want to cherish. This is why I stopped and stepped back a moment, after I had reached the door, and turned to take a last loving look at the peaceful streets veiled in mauve twilight, much like a Moslem woman in *purdah*. Enchanting! I could hardly see the turmoil within the brightly lighted National Hotel. But I was still a part of the magic of that twilight in Jerusalem. Never again would this moment return except in my memory. So I hoped, by pausing, to cinch it someway for all time to come. Dwight, however, felt no sense of reluctance about entering the hotel. He was hungry! And time was going fast! So, adjusting his camera bag, he swung on by me and opened the door, holding it for me to enter. There was nothing to do but follow.

Some of our Tour Members had already eaten dinner and were milling around in the lobby, talking loudly. They seemed upset about something. But this was nothing new, so I attempted to get past them to the elevator. But the desk clerk called us, and in a flat, impervious manner told us the sad news—we were to leave *very early* the next morning for Egypt! All our large luggage must be packed and in the lobby for loading before 7.30, and we, and our hand luggage, were to be ready to take off no later than 9.30. Having told us, he quickly turned away to some other task, leaving us standing stunned and staring. Then Dwight recovered enough to thank him and guide me toward the waiting elevator. But I didn't "come to" until we stepped out on our floor. Then I began to bemoan our fate: "7.30! Why, that's just ridiculous! We are supposed to have another full day in Jordan. This way we won't get to go to Gethsemane, the Virgin's Fountain, or any of the places we have missed," I continued, walking rapidly down the hall at the same time. Then, thanks to the Spirit who quickly reined me, I realized that I was talking and acting just like a "tourist" and not like a "pilgrim." So I said no more, but tears were overflowing as we entered our room.

Dwight took it much better than I. He was not appalled as was I, at the thought of being all packed up by that awful hour. He quickly set about cleaning up for dinner, not realizing that I had lost my appetite. Oh for a quiet half-hour or even a quarter-hour with the Lord! But, praise God, He has taught us to be "instant in prayer." So, a quick dart to Him brought a prompt answer. The Spirit impressed me that I must not allow this development to distract me and spoil the precious remaining *PRESENT*. And I must practice "presence of mind." This evening was the Lord's, and I must redeem every possible moment of it. So I quickly began to praise and thank Him, and to clean up for dinner. We had no time for baths—just a quick change and we were on our way. I had hoped we might dine alone again, but a few stragglers remained. They were all displeased about leaving, and the chorus of complaints was all around us. They all agreed that we had been promised three full days in Jordan, and had had but two. I was indeed thankful that the Lord had already met me and given grace. It would have been a terrible waste, as I see it now, to have marred that last sweet dinner hour in Jordan. How kind and attentive the waiters were! How good the food! How near the Lord!

We dared not linger, however, for the Mattars would soon be arriving. Thinking still of Souad, I tried once more to locate her by phone. But to no avail. Then I got a wrap and my harp case, thinking we should be ready to go at once with our friends when they arrived. It was already growing late, it seemed to me. While we waited in the lobby, with mounting tension in my heart, I tried to praise and look to the Spirit. Several came to question me about our orphan and the home in Bethlehem. One or two expressed a desire to adopt one too. So the waiting time was filled with talk about the Lord and His interests. But still the Mattars tarried! I was truly uneasy by the time I saw them coming toward us. However, they were quite relaxed and sat down in the comfortable chairs of the lobby for a leisurely visit—displaying a gracious but sometimes frustrating Arabic quality of relaxed indifference to *time*.

The talk was all about natural things—our trip, our country, the plane, the people, etc. I grew more and more concerned. Apparently they had no particular desire to take us to the Mount of Olives Chapel, after all. Why, oh why? It was a mystery to me. But I knew that I *had* to go. It was one of those times when I felt that desperate measures must be taken if necessary, in order to obey the Lord. But first I tried gentle hints, such as: "Did they think I would need a heavier wrap up on the Mount? Was the road to the Chapel very steep? There must be a beautiful view up there by night." All to no avail! Sister Mattar said that, "Yes, my wrap was sufficient; no, the road wasn't very long or steep; and yes, the view was lovely." And the conversation continued steadily on. Dwight, of course, was entirely unaware of my sense of expediency, and was enjoying such good conversationalists. So, something more drastic had to be done. But what? About then, the Lord must have intervened, for Brother Mattar suddenly noticed the harp case, which I had purposely put conspicuously on a coffee table in our midst. He asked what it was. This gave me a chance to answer and tell him that I was taking it up to the Chapel so that I might sing a special song the Lord had given me about it, a sort of dedication song. (The Chapel was supposed to have been dedicated during the time we were in Jerusalem, but delay in the building process has caused a postponement.) Since I couldn't be there for the service, I would sing the song tonight, I explained. This did not arouse any interest whatever. Both of them looked at me rather blankly, and then started talking about some other matter. It appeared to be a deliberate snub. But I realized that more was involved—there was a definite opposition to the Holy Spirit, such as we had all encountered during the afternoon service. Good old Dwight then quickly rose to the occasion. He looked at his watch and got up saying very positively: "Well, look at the time, we must go at once to the Chapel, because we have to get back and pack tonight for an early morning flight to Egypt." This produced the desired effect, and we all made our way to their car.

When I saw it, I wondered if it were partially the cause of their reluctance to take us up the mount. It was an antique indeed! And getting it started proved to be quite an operation. But finally it charged its way through the quiet streets and on to a road that wound around for a few blocks. I was surprised to note how dark the city was, even at this hour. They do not have the ample street and building lighting which we Americans take for granted. I have heard that when the King comes to Jerusalem, and on special Feast times, they do light up even the old walled city with brilliant strings of colored lights. But ordinarily it is very dimly lighted indeed.

When we got out of the car it was too dark to see even the roadway, and the only light we had was a flashlight which Sister Mattar carried, as she led the way. Why we had forgotten ours, I did not know, and was tempted to be a bit disgusted with us for doing so. We followed her to a rather steep cement stairway, and then on across a lot which had a most uneven terrain. Dwight tried to hold on to me firmly, and I wondered why I felt so shaky all of the sudden. I might have known! The house and chapel were dimly lighted also. Only two people were staying in it at the time. Perhaps I should explain that the house has been there for a number of years. When Brother Mattar obtained the property, by a near miracle—he decided to build the chapel on to the front of the house. It has windows all around and overlooks the city in an auspicious way. It is quite a bit higher on the Mount than the Garden of Gethsemane, and a little to the left of it, as you look up. In fact, it is near the summit of the Mount, and not too far from the Chapel of the Ascension.

Jesus surely must have walked over all this dear old Mount of Olives. In those days the city itself lay some forty to sixty feet lower than it does now, making the mountain appear higher. But it would not take long to walk about it, so I am sure Jesus must have prayed or meditated in many different spots, by various rocks and trees. The city He knew lies buried, and few of the old landmarks remain. But the Mount of Olives has changed but little. The thought that we might be walking over ground sanctified by His own steps was overwhelming. I felt that I was going to swoon with love for Him—but dared not. By now I knew that I was undertaking a dangerous and important mission, the full significance of which was partially concealed from me. So I must be alert and watchful. There was no timeout for swooning.

We met the guests, missionaries from Africa, and sensed that there was some sort of tension between them and the Mattars. Rather reluctantly Brother Mattar invited them to join us in the chapel. Chairs were brought in and the doors were left open, so that the lights of the house could shine in—there being no electric wiring yet in the chapel. It was a large and very plain room with arched windows, and as yet it had no furnishings. I longed to walk about it, lift my hands in praise and really thank God for it. I tried to recall the exquisite joy some of us have felt when we learned, through Dr. McKoy, that this property had been obtained and would be used for a Christian Protestant Chapel on the Mount. It seemed almost unbelievable that we could have part in providing a house for the Lord on the very Mountain from which He ascended and to which He shall return! The Spirit spoke clearly to us then that, having provided Him a house (and houses) on this Mount, we were now to have this privilege in the Holy City on the Holy Mount. Wonderful! I could picture all of you there with me, moving about the room, singing, shouting and praising. A few of you were kneeling in quiet adoration.

My reverie was interrupted by Sister Mattar's gracious hospitality. We have learned that when one is received into an Arab home, refreshments are served at once, to assure you of welcome. Usually Turkish coffee and little sweetmeats are the fare. But, in this case, an orange soft drink and some cookies were used instead. I was not hungry at all. But I was touched at their kindness. The room grew so hot that the large front door was opened, and the wind rushing in was surprisingly cool, after such a hot day. It seemed to beckon me to come out and walk on the Mount. Here was another moment of agony and joy. I was finally on the Mount and in the Chapel! But I had no liberty to do the things I desired. To walk and sit by night on the Mount was a privilege I had long desired. During the time of our Gethsemane watch—many years ago now—the Spirit took me to this Mount night after night. How very real it became! Yet I yearned to walk there in the body also. When the refreshments were ended, and not wishing to be rude, but torn by this urge, I asked them all to excuse me for a few moments while I viewed the city from out of doors. Those few moments were priceless!

As I stepped out into the garden, an indescribable atmosphere was all around me. The City and the Mount were filled with history and mystery! Yet there was something more—expectancy! Does this dear old mountain know that those beautiful feet that once blessed it are soon to walk again upon its brow? Does it sense that there will be a great cleaving of the earth, and it will divide as did the veil of the Temple so long ago? Surely the angels keep watch over the City and the Mount—and were present that night! I was amazed when I realized that I felt the very same atmosphere as I had when I was given the Zechariah song—and visited the City and Mount by the Spirit. Wonderful! Awesome!

This interval was all too brief. My hostess was already seeking me, not understanding why I had withdrawn. It would be a breach of love for me not to return to the circle. So I had to turn away and leave behind this marvelous moment also. Almost rapt with the wonder and pain of it all, I felt it was time to sing. But how to interrupt the conversation, now in full swing, I knew not. Again the Lord helped. The missionary brother asked if we were going to have some music. He seemed to expect a sort of hymn song, and was obviously disappointed when I took up the harp and explained about the dedication song. I have seldom sung to a more unresponsive audience—if one thinks of the earthly aspect. But who knows how many of the heavenly ones were listening to this offering from the ardent heart of an earthling? As I sang, I again felt all of you with me and that I was representing you before the Lord and the heavenly host. The melody is in the minor and is worshipful and, to me, nostalgic.

> Here on this mountain, this memorial mountain,
> This mountain called Olivet,
> Where Jesus often walked and sometimes talked,
> Speaking words men will never forget.

> Dear Mount of Olives,
> Sacred Mount of Olives,
> May Jesus be honored here!
> May His voice be heard
> And hearts be stirred,
> Feeling His presence near!

> Here, where He often came to pray
> At the close of a weary, heart-breaking day,
> We come to lift up our hearts in His praise,
> And to offer a hymn to His grace.

The Holy Spirit came upon me in a remarkable way, considering the circumstances. And in the silence that followed the song, I spoke of how the Lord had quickened to us the prophecies of Zechariah. I then sang my Zechariah Song, which begins: "The Spirit moved upon me in a strange mysterious way, and I seemed to be transported back to Zachariah's day." I never sing this song without feeling my spine tingle! And that night was no exception. But only one of the little circle seemed at all moved upon. The others waited a polite moment, and then resumed their conversation.

I know you will all share with me the burden I felt that night—and still feel— that this Mount of Olives Chapel shall become a place where the blessed Holy Spirit may be recognized in all His ministries—not just in proclaiming the Gospel. Surely our Lord desires and deserves such a place on the Mountain where He agonized and sweat drops of blood. I felt a little of His agony that night, as I realized how well-intentioned and yet how limited these dear ones were. Later I heard the brother say to Dwight, "We are Baptists and NOT Pentecostal. But we go along with them." This so-called "going along" usually means mere toleration, I have found, and hidden, if not open, opposition.

As we were taken on a tour of the house, the burden increased. The enemy was permitted to attack me in a most cruel and sudden way. And his dart found its mark with deadly accuracy! I know the ones involved had no realization of how he was working. But, praise God, we are not ignorant of his devices and ways. We know our battle is not with flesh and blood—yet we discern the vessels he uses. Later, in thinking of this unexpected attack, coming after so significant a day, I was reminded of the Gethsemane song I so lived in—during the time of the "Watch." The lines, "I bless the cup, as I lift it up, tonight in dark Gethsemane," came to mind. And I realized that it was indeed in order that on the Mount of Christ's agony, I too shared that cup to a small degree.

As we left the Chapel the night seemed yet darker, even the sky appeared overcast. We stumbled along toward the car. When we came to the stairway, Sister Mattar led the way, again carrying the flashlight, and I followed directly behind. Just before we reached the top, she suddenly whirled around and struck me a sharp blow on the head with the light. So sudden and surprising was this that it almost caused me to fall backward, but Dwight caught me. In this strange interval time seemed to stand still for a moment, and Jesus suddenly manifested Himself in me, stifling my cry of surprise and pain and speaking directly through me. I opened my mouth and heard a voice very different from my own speaking! This was even more surprising than the blow, for I was feeling utterly crushed and He seemed an eternity away. But He was right at hand! Grace flowed in place of tears. I practically embraced the sister, who was apologizing, and heard this sweet Voice assuring her not to fear, that I was not hurt. Yet my head was cut open and bleeding! I know now, even better than at that moment, that had Jesus not stepped in, I might have cried, or said something negative, for I was already shaken to my soul, and realized that this was indeed the hand of the enemy. All my testimony and anointing might have been undermined had I not had grace in this critical test. Bless Jesus, He was faithful to step in at just the right moment!

By the time we got to their home—which is located in the Garden of the Tomb where Dwight and he were to discuss building design and other business matters, I had recovered enough to have presence of mind. I had stanched the flow of blood with my handkerchief, and since my hair is thick, the gash did not even show in the dim lights. For another hour I sat, attempting to be polite and responsive. But all the while the Spirit was dealing and revealing things to me. I can assure you that there is a great need of prayer for these dear people, this Chapel and Garden, and for the city itself. With all the religiosity a clear Holy Ghost witness and ministry is sadly lacking throughout the area. Under this burden, and after the extreme experiences of the day, I was scarcely able to hold up my head. I was tempted to get upset about being kept so long, for it was by now near midnight, and our laundry was not done, nor our bags packed. I was near the point of collapse when the discussion finally ended, and the plans were given to Dwight for his further designing. I felt to leave an offering from us all for the Chapel, in spite of all the negative things that had happened. This is ground to *hold*, not *abandon*! Brother Mattar had already given us bookmarks and flower cards for our Company, and sent greetings to all.

At last the goodbyes were said and we were walking out through the garden toward the gate. What a contrast from our entrance through that gate earlier in the day! I was as eager to go as I had been to enter, but I could hardly put one foot ahead of the other, and my head was throbbing painfully. The Brother had offered to drive us to the hotel. But the car refused to start. So we tried pushing. After a few minutes, we all realized that it was useless. He offered to call a cab, but Dwight was still feeling "high" and excited about the plans, so he told him that we didn't need one, we would walk to the hotel. I was surprised to find myself agreeing. But I was eager to be on our way. As exhausted as I was, I was moved to tears in gratitude that once more I could walk through the blessed streets of the city—and this time by night. The appearance of the old walls, the Damascus Gate, and the towers and domes, seemed even more ancient and awesome at midnight than during the day. I felt as though time had turned backward for a few moments. "I walked in Old Jerusalem by night, as he (Zechariah) did then; and I saw the angels all around, encamping there again."

It was true! It was real! But only for about a dozen steps. Then fear struck me. It was dark and we were "rich Americans"—behind any wall or building an enemy might lurk! (After all, I had met the enemy all during the evening.) I was immediately ashamed of my fears, of course, and began to praise. How awful to mar even one moment in the city with negative thoughts! As I praised, I felt at ease and at home. We passed a laughing group of Arabs who smiled in a friendly fashion. We saw a policeman who gave us only a brief salute. Then in silence, we walked the last block or two, each of us too deeply moved for words. By tomorrow night we would likely be far from Jerusalem. This was the beginning of farewell. We were both in tears when we entered the hotel.

For once the desk clerk lost his composure and leaned back, incredulously staring at us. I suppose all the rest of our tour companions had retired fairly early to be prepared for the next day. Doubtless he wondered what on earth we had been doing on the streets of sleepy Jerusalem at midnight. But then we had been only partially *on earth*! I had felt the forces of both hell and heaven. I had walked almost two thousand years ago with Jesus on the Mount, as well as centuries before His time, with Zechariah, through the quiet streets. Angels had hovered near!

We picked our key from the still astonished clerk and cheerfully told him good night. He came to his senses and reminded us that at seven-thirty sharp all our large luggage would be loaded on the bus and taken to the airport. We assured him ours would be ready. The elevator creaked its way to our floor and we quietly walked down the oriental carpeted hallway, just as we had done three nights before, at about the same hour. The sight of our room shocked me back into reality. Nothing was washed, nothing was packed! It was all a jumbled mess of belongings. But neither of us was equal to doing another thing. So we agreed to get up early in the morning and do our packing. I walked out on the balcony and took time for a last, lingering look at the Mount of Olives by night. A few lights still glowed. I had ascended the Mount twice that day. I had visited the Chapel. Mission accomplished! What did it matter if I had been wounded in the battle, both physically and spiritually. Tomorrow would be our third day in the Holy City. Maybe it would be a glorious resurrection day!

Note: The Mount of Olives Chapel has been dedicated and is in use. The Gospel is preached there on Sunday nights. It is available for prayer and meditation during the day. The guestrooms provide a place for visitors. And clean Arab style meals are being served. At present they are considering a large "prophet's chamber" on the third floor. This would not only overlook the city, but also all the countryside even to the Dead Sea. We are having a part in the construction of this room, as the Lord has led. The Mattars have asked me to represent you on the Board of Trustees, and are incorporating as a nonprofit religious organization. They have written that this is our home in Jerusalem, to which we shall always be welcome if any of us can come on a visit. But we are more concerned about it being a true home for Jesus and the Holy Spirit. So let us keep praying and believing.

FAREWELL TO JORDAN

D awn has a way of arriving unreasonably early in Jerusalem. Perhaps it is because the prophet once cried out, "I said to the dawn, 'be swift!'" Or, maybe because the Psalmist sang with such ardor, "I will awaken the dawn!" In any case haste seemed most inconsiderate to me, especially on this, our third morning in Jerusalem. Dwight, on the other hand, was inclined to hurry it along. He has a way of awakening very early every morning, and even earlier on days when he has some special activity in mind. Thus it came about that before there was any eastern light in the sky, Dwight began padding around the crowded room, stumbling over this and that, hunting for his flashlight and, finally, ending up standing out on the balcony looking out upon the still sleeping city. That is just what I longed to do too—sleep! So I resorted to subterfuge and pretended that I didn't hear a thing. I didn't stir a muscle until I finally heard him get back into his bed and yawn. How I hoped he would go back to sleep! Surely it was much too early to get up, for the hotel was absolutely quiet. When I finally ventured to stir a little, I discovered that I hurt all over and not just in spots. My head was still sore and throbbing a little, my body had sore spots well scattered, and I felt as though I had been quite soundly whipped. My feet were blistered and protesting loudly about the cruel way I had been abusing them. I sighed deeply, feeling very much like Mrs. Job. This evidently wasn't going to be that glorious "resurrection" morning I had hoped for! At least not in a physical way.

Suddenly a temptation assailed me to "push the panic button." If only there were some way for me to get out of this "space capsule" in which I was hurtling at such speed, and be back safe at home in my own bed! We had toured only two days after our arrival, and already I was completely exhausted. Yet almost a month of such touring lay ahead of me! Doubts assailed me. I realized that it was too late to be ejected from my "capsule." Why had I ever imagined that I could stand up under such a grueling test? Like most of you, I can arise to a crisis and bear up for a reasonable

201

time. But after a few days a letting-down period is sure to come. And on this trip no intervals had been allowed for such. On the verge of tears, I suddenly realized that resorting to self-pity and negativism would but add to my miseries—and allow the enemy just the ground, just the opportunity, he was seeking. So I turned deliberately to praise and thanksgiving. Of course, I was very quiet about it, hoping Dwight was by now fast asleep. He wasn't! All of a sudden he sprang out of bed and announced that it was time for us to get up and pack, if we were to make the deadline. The shock of this was so great that I again started to cry. After all, I reasoned, during the last four nights I had slept only about 15 hours at the most—instead of the normal 32. And during the days I had been at least twice as active as normally. What could be expected but fatigue?

How thankful I was and am for the years of training by the Spirit and the Word. One just does not give in to such fears and thoughts, nor dwell on mere human possibilities. We serve the God of the impossible, so is it unreasonable that sometimes we are called upon to prove that? I must do as in every crisis—praise Him! Thank Him! Affirm my faith in Him! Rely on His strength alone! I must greet my husband joyfully and get out of bed promptly, acting cheerful and confident. Mornings are never easy for me at best. So I have to practice a cheerfulness I do not feel every morning. It came in handy now! (Of course evenings are a different matter. I am often still bright at midnight!) I began to hum, though with a quaver, for I was truly in much physical misery. I was glad that by now others were stirring, running water and talking. We could take our quick baths and get to the packing. And we did.

"Say that you are well and all is well with you, and God will hear the words and make them true." Affirmation plays a part. So I tried to be positive in my thoughts, as I sang a little and praised. But I simply could not shake off the sadness I felt over leaving, even though I did manage to ignore "Sister Body's" interior moanings and protestings. (Years of practice at this paid off too.) About this time another attack came upon me—a fear of going to Egypt, a dread of all I would have to go through before I could again lie down to rest. And, even worse, the sudden realization that I would have to fly again! Having safely landed out of the air five times, my affection for the terra firma was greater than ever, and I had no desire to leave it again. For a few moments I staggered about in my thoughts, wondering if there were not some way to stay in Jordan and skip Egypt. We could rejoin our party in Jerusalem, Israel. We could find hospitality with the Mattars. We could visit all the places we had missed—the "Hill of Paradise" and the site of the once beautiful Gardens and wells of Solomon; and dear Hebron, where David was first crowned, and Abraham, Sarah and others lie buried; (this was Caleb's land too, where he and Joshua first spied out the land); the Virgin's Fountain, which Dr. Field has said is the most important landmark in Jerusalem—still flowing; and dear Ramah, also called Ramathaim-Zophim, the home of Hannah, where Samuel was born, lived and died, and where Deborah lived also. Perhaps we could go further north to Bethel, from whence a marvelous view of the land is possible—and even to Samaria. In Jerusalem there were so many places we had not even glimpsed: the famous Rockefeller Archeological Museum, the Zion Covenant, and other places of interest. I was indeed torn as I underwent this attack, for I was not sure if the Lord were trying to show me that He desired me to stay, or if it was my own desires and fears that were impelling me. A struggle like this can consume a lot of time and strength and I had none of either to spare. So I had to settle it and overcome quickly. I did! I recalled that the Lord had shown me in advance that I was to accept THE TOUR as His will and not seek to follow my own inclinations at any time. I was to discipline myself to cooperate fully. The Tour was now preparing for flight to Egypt, and this is just what I should be doing too. I resolutely turned away from this fear and temptation and began to pack and praise. Yes, it truly proved to be His will for me to follow my Lord, and Joseph, Jacob, Abraham and the Israelites to Egypt. Praise God that grace was given!

By this time Dwight had spread the beds with all our belongings—mine on one, his on the other. He is an expert packer. How I wished I had practiced more before coming, for on tours it is good to pack your bags the same way each time, so that you can locate things even in the dark, if need be. It seemed incredible that all this pile of things had to go into our limited luggage. But I began valiantly. Dwight gave me his reading glasses again, and everything immediately rose up and fairly smote me in the eyes. I couldn't see for seeing! However, when I removed them, everything fell back and blurred together. The light was very dim, and I was soon frustrated and confused for, in addition to what we had brought with us, there was a large pile of things we had purchased or collected—gifts, souvenirs, maps, booklets etc. We tried dividing this between us, and shifting things around to make more room. But, after a few efforts to cram these in, we realized that it was useless. We just couldn't go through this repeatedly, each time we packed. So Dwight decided to get a carton and pack all these extras for shipping. It seemed a simple solution, but proved to be quite involved. By the time he got a bellboy and a box and we finally got it and our large bags packed, we were really exercised. But we did make it by 7.00 o'clock, and sighed with relief that we could now go to breakfast.

We were resolved to be in a thankful, cheerful attitude, for we knew how the atmosphere was sure to be. It seemed that almost no one wanted to leave Jerusalem and go back to Egypt. We stayed resolutely positive throughout the meal, and irritated our neighbors by conflicting with the complainings. Then we said our thanks and goodbyes to the very friendly waiters, and gave our tip to the headman. (Although our tour supposedly covered all tips, we had learned that this meant that many who served us got little or nothing.) Hastening back to our rooms, we set about packing our hand luggage. Then we congratulated ourselves that we had made it, and had left nothing behind—meanwhile overlooking my especially useful folding plastic clothes hangers, needed particularly for drip-drying! We said a few words of thanks to the Lord, stood for the last time on our balcony and looked at the city. Then, not without tears, we picked up our things and headed for the desk. Of course the dear attendants rushed to help us and almost wept to see us go—especially the one who had escorted me about the roof and taught me Arabic. Two of them begged Dwight to have his picture taken with them. So it was shot. We had already given them their tips, so this show of affection seemed genuine. And I know that we truly loved them. May Jesus save them and their families!

Our next project was to get our package mailed. Ahman offered to speed this up and led us down the street to a shipping broker, explaining that all packages shipped by boat have to be sewed into cloth. Also I had to make out a complete list of what I had bought and its value. If this didn't prove exacting! There was much talk both in Arabic and English, and quite a waving of hands and gesticulating. Since any two conversing in Arabic sound like they are having a hot argument, I grew increasingly disturbed. I hoped everything was legal and that our highly valued (to us) things would eventually reach us. (They did.) I had no fear of missing our bus, because

Ahman was with us. When we did return to the hotel, we found everyone in front or in the lobby. It was time for departure. I stole a few minutes to write some cards, and left them, with stamp money for mailing. (Most of them never arrived.) Still no bus! To escape all the confusion and grumbling, we decided to take a last short walk down the delightful street. Soon we ran into George, the other guide. He was trying to escape the grumbling too, I think. He seemed relieved when he saw that we were joyful and friendly. Noticing an Arab boy carrying a tray of Turkish coffee—a common sight on the streets—I chanced to remark that I hated to leave without having tasted it. Whereupon George gallantly invited us to come into a small shop nearby and be seated. He said that I should be served promptly with my coffee; and I didn't need to be urged. Dwight refused, since he never drinks coffee. While George and I sipped the tiny cups of very strong, syrupy, sweet and spiced coffee, I felt a love for him, as I had when he escorted me in Bethlehem. May he find true salvation in Christ! We assured him that we were eager to visit Jordan again, and we mentioned how sorry we were to have to leave without going back to Gethsemane for a visit. He told us that the bus had been delayed, (which they knew all along, but didn't tell us. I think they just rushed us out of the rooms to give the maids time to clean up before the next group arrived.) He assured us that if we wanted to call a taxi, there was still time for a short visit to Gethsemane. We jumped with delight and he promptly called one. He told us to give the driver only a dollar and to ask him to wait. He said we could take a half hour safely.

Resurrection set in at this point! I was revived as if I had slept all night! And I was filled with joy. Had Dwight and I not maintained a joyful, positive, right attitude, this would never have happened, for we would have been back in the hotel milling around with the rest of the complainers. Feeling very out of place with them, we had gone on our way, and the Lord really met us in George and then again in the Garden—for we were soon walking there. It was heavenly to be able to go in quietly alone. There were only a few visitors on this calm Monday morning. One group from Sweden were over in a corner, in a circle, singing softly and later reading from the Word and praying—how worshipful, even blissful, their faces looked! All tours aren't alike! While Dwight took pictures, I walked up and down the paths, lingered under the ancient trees, which still bear fruit, and noted all the lovely flowers. The holy-hocks (as they were first called) were in full bloom, and they are native to the Holy Land. Many were much taller than I, and Dwight took a lovely colored picture which, alas, did not turn out right. I looked across the Valley of the Kidron, and gazed long at the Golden Gate and the Dome of the Rock, thinking how Jesus saw the Gate and the Temple beyond, as He often prayed here or close by.

Then I had a longing to go again into the Church of the Nations. It was almost empty, and I lingered again at the Rock of the Agony, and walked up and down beneath the tranquil heavenly light softly filtering down through the vari-colored alabaster windows. I rejoice to relate that our beloved Lord did draw very near and spoke to me there and in the Garden. His words were personal and most comforting. I also felt His desire that I share all that I possibly could of the trip with each of you, for He had great desire for you to know and love and visit His land too. I realize that perhaps in some cases this visiting may be only in the Spirit. Yet, for some, it may become literal too. I had a sweet assurance that I would return in the body, as well as in the Spirit, to this dear place, and that He desired this greatly.

I was uplifted! I was renewed! I was now eager to follow Him to the darkness of Egypt's land. And Dwight, of course, was as happy as a lark, taking pictures and noticing everything with much interest. When we returned to the waiting taxi, I looked longingly at the little church that covers Mary's tomb, which is just across the little road. I had hoped to visit it, but could only pass by and honor her, as I did. I took a last loving look at the pathway leading up the Mount from the Garden. "Next time I will follow it," I promised the Lord, for it is quite certain to be the age-old way that Jesus took. Each Palm Sunday a great procession forms at Bethpage at the top of the Mount and follows this path down into the city, carrying palms, banners and other tokens. There is singing and praise. And tourists too may join in. Wouldn't that be a thing of glory!

Our return to the hotel stirred up more disapproval toward us. Where had we been? Why did we risk missing the bus by running off? What a furor! When Dwight explained that George had arranged it, at our request, some were even angry. They said that they too could have run around some, had they known. Why hadn't they been told too? It was most embarrassing. But about this time, fortunately, Ahman announced the arrival of the bus, and we all made a grand rush toward it, as usual, each hoping for a good seat, I guess. The Airport lies to the north of Jerusalem, and soon we were passing by the Mount of Olives, and following the upper Kidron Valley. On our right we could look up and see famous Mt. Scopus, where the old Hebrew University stands. This place is still in dispute and cannot be used. Gibeah of Benjamin (1 Sam. 21:1) lies over to the right of Mt. Scopus. It is now known as Tell el-Ful. Not far from here lies the village of Ramah (Er-Ram now), Samuel's village. Left from Er-Ram is a paved road leading to Ej-Jib (He comes!), the biblical Gibeon. (It is interesting that archaeologists have now uncovered a large chasm carved in bedrock there, which is identified with the Pool of Gibeon. (2 Sam. 2:13) Not too far west is the town of El Qubeibeh, thought to be the biblical Emmaus. Our guide tried to describe these various biblical places nearby, even though from the bus we could not see them clearly.

As we turned toward the airport, we were surprised to see how very small it was. I wondered how any large plane, let alone a jet, could possibly land there. There seemed to be only one main runway, and we soon learned that it actually runs across the road on which we were traveling. Gates were let down, and we had to wait for a plane to come in. This aroused a sense of amusement, I am sorry to say, and several began to make jokes about it. I immediately sensed the pain this caused our dear guides. After a short wait, during which they told us that our plane had been delayed, and that they were going to take us on a "picnic" to Ramallah (joy! joy!)—a plane did arrive. What a relief that it made a safe landing, it looked precarious! But still the gate stayed down. Another wait! More joking! Then a bright yellow plane, not much larger than a Piper Cub, set down gracefully and taxied to a stop like a tropical bird. "It is the King!" Ahman and George shouted together. "That is his little personal plane." Sure enough, it bore the King's insignia! You can imagine the effect this had on me, quite opposite to the one it had on the men. They began to laugh and make jokes about a king flying around in a little plane like that—why anyone in America could have a plane that size, etc. But I knew and loved Hussein, bless him! And felt that he was very important to God, to Jordan and to us. And I had learned also that he is an expert pilot, not only of little planes, but of jets as well. And of course he does have some jets, but uses this little plane for short hops. With effort I restrained myself from reproving my brethren. The gates lifted and we were now climbing up in the hills approaching beautiful Ramallah. This town was almost destroyed by earthquake in 1927, and has been rebuilt with houses of stone. It is a gardenlike spot almost 3,000 ft. high. Because of its cooler climate, it has become a resort-like area in summer for the sun-blasted Jordanians who can afford to go there. It is a Christian town, delightfully clean, and much enjoyed by tourists too. We found it pleasant to be driven all around its lovely hills. We saw very fine looking Arab children in the streets and the people here looked prosperous and very happy.

After circling around for a second view, we turned back toward Jerusalem, passing through another garden spot—a little town called El-Bireh (the well or cistern.) Some believe that this is Beeroth. (Josh. 9:17; 2 Sam. 4:2,3) Traditionally this was the first resting-place for caravans proceeding from Jerusalem to Galilee. Thus El-Bireh is most probably the point where Mary and Joseph discovered that Jesus was missing, and turned back to search for Him. Our guides talked about the famous incident, and seemed to find much pleasure as well as pride in showing us this green, well-watered, more beautiful part of Jordan. How thankful I was to the Lord and to our guides for letting us have this delightful side-trip, instead of cooping us up waiting in the hot airport.

When the bus turned back toward the Airport, several clamored about where our "picnic" would be served. (During the rest of the tour we sometimes laughingly referred to this novelty of "a picnic without food.") The guides assured us that we would be given lunch on the planes. And by now we were most eager to eat for it was high noon, which in Jerusalem seems very high indeed! We had already generously tipped our guides in a general fund, and they seemed most gracious and kind. They need not had given us this extra trip. Arabs can be most charming! We parted with them reluctantly, assuring them we would return.

Then we jammed into the building, and with dismay saw that part of our Tour, which was scheduled to leave ahead of us, was still waiting for a plane. They were tired, hungry and disgruntled. So our prospects looked very poor indeed. We had the inevitable soft drinks and a few consoled themselves with candy bars. Then we just stood and waited, for there were no vacant seats. The plane being readied looked rather small and old. And some of the men began to say that it would never get off the ground, let alone across that desert. I was already feeling mounting reluctance to emplane—and this added to my fears. I had to exercise myself to keep outwardly calm and cheerful. How grateful I was to some woman who told me that these planes had been made in America and were sort of a "workhorse" type—having been used for many years all over the world. She seemed to think them quite safe. But, of course, our faith was in the Lord anyway, or should have been.

After a short while the other section were called and rushed out to get on the plane. Their manners were very bad, as most of us observed. And yet I knew that we too acted likewise. It was apparent that everyone wanted to pick his own seat. I wished they would just assign them. The men shoved the women, and some of the older women were almost knocked over in the rush, and gradually learned to wait until toward the last. It was shameful, I thought, for anyone, let alone Christians, to act so rudely. After this group departed, some of us could sit down for a while. It was very hot, and we wished they would let us wait outside. About an hour elapsed before our plane was taxied into position. We learned later that due to a mix-up in plans, this crew too had worked hours overtime. Yet the little Arab stewardesses were alert and gracious as they stood erect to greet us. When the gates were opened there was a veritable stampede, and I caught myself being pressed into running—so I got to one side and slowed down. Christine was running too, and suddenly lost her shoe. She stopped to get it and burst into almost hysterical laughter. When I reached her she was gasping, as she laughed, "Oh isn't this just ridiculous! Here we all are running for our lives, just as though the plane was going to go at any moment and leave us." She kept laughing and panting, saying how awful and yet how funny we must look to the Jordanians, pushing each other aside in crude American manner. She and I had kept laughing as we walked slowly to the plane in a ladylike manner. "Hereafter," she said, "I am going to get on planes in a Christian way, even if I get the worst seat." I agreed. How dear of the Lord to give us grace to exit laughing, when we both really wanted to cry!

We had poor seats, as Dwight usually depended on me to go ahead and get them, thinking ladies should be first—an old-fashioned idea not many other brethren shared. We were right by the wing and the motors—and were they ever loud! So we could see little and hear lots. I was disappointed, for I had wanted to have a last, broad view of the beloved city, and then a glimpse of all the precious land we would pass over. But, as it happened, this would have been impossible anyway, for after we made a short run, and to everyone's relief actually got off the ground, we rose quite swiftly and were at once swallowed up by the clouds. There was nothing to see—but oh so much to feel! This was farewell to Old Jerusalem. It seemed to have been swallowed up by time, and we by the clouds. All spent now, the priceless hours and days! Much of the Land where Jesus walked was now behind me—and my heart was left behind too, even more certainly than the little tokens I had painstakingly hidden at every place where I had walked. I had set up a lot of unseen altars for all of us in that dear land. And neither I nor it will ever be quite the same because of our First Missionary Journey.

I am writing this at the anniversary time of our departure last year. It seems incredible that I have been home now for about eleven months and have been writing about the trip off and on during all that time. Yet only in this portion do I finish telling you about Jordan. How could so much happen in three days and nights that it would take eleven months to tell about it? Yet it is true! Already I have written an estimated 36,000 words. Enough for a good sized book! Each portion takes much thought, meditation, prayer and waiting for the leading of the Lord. I also have to look at various facts about places visited, to refresh my memory. And each writing is like a going again, for I find myself seeing many things about the trip I had overlooked, as I attempt to recreate it for you. I praise God that the Holy Spirit does anoint and assist, otherwise, I should dread to try to continue. But I trust to go on and complete the record—no matter how long it takes. So pray with me toward that end.

I have been led to review during this month all that I have written about the trip to date. And I am happy that several of you told me that you were rereading it all and finding many new things in it—or new impressions—in so doing. So rise up and come away with me again, on the wings of the Dove! The Land is calling!

WAY DOWN IN EGYPT'S LAND

F
ew of the songs I learned during my years in Grammar School impressed
me enough to remain in my conscious memory. But there was one song
whose minor strain deeply moved me as a child, and has lived in my heart
and memory all during the more than forty-eight intervening years since I first
heard it:

> When Israel was in Egypt's land,—
> Let my people go!
> Oppressed so hard they could not stand,— ·
> Let my people go!
> Go down, Moses!
> Way down in Egypt's land;
> Tell old Pharaoh,
> Let my people go!

"Way down in Egypt's land" became a very real place to me, not just a storybook
land. And even in my childish heart there was a desire to go to that land, and a
certain awareness that someday I would visit it. Perhaps my response to this song
was inspired by a special interest my mother had in its pyramids . . . especially in
the Great Pyramid. I can still recall sitting solemnly on a hard, high-backed chair in
the home of our landlady and neighbor, trying to listen and understand the lessons
a white-bearded Bible teacher who was expounding about Egypt and the pyramids,
to which he seemed to attach such significance that his face glowed with rosy light
and his eyes behind his "specs" fairly danced with joy. The words he spoke were as
far over my mind as the chair-back was over my head. But the rosy glow and his
dancing eyes fascinated me. Apparently he and those present, including my sweet
mother, believed that this pyramid was indeed "an altar to the Lord in the midst

213

of the land of Egypt," and had great prophetic as well as historic significance. Doubtless too I heard often of Egypt in early Sunday School lessons, and formed many mental pictures of Joseph and the Israelites, Moses, and even of Jesus and His parents in Egypt. Then, too, while I was still quite young the first production of the Ten Commandments as a motion picture was made. And how deeply this spoke to my heart and mind, only the Lord understands. I seemed to enter into the terrible sufferings of Israel. I heard the old spiritual ringing loudly in my own soul, "Go down, Moses, way down in Egypt's land . . . . tell old Pharaoh, let my people go!" I saw with my own eyes the Red Sea turned back and the people pass over on dry land. How real, how wonderful, how great this picture made God seem!

All this was more than forty years ago! And now the time had come, on this date in May 1961, at the time of the new moon of the month Sivan, that I was being borne in person to a predestined appointment with God "way down in Egypt's land." Why was I so reluctant to keep this tryst? Why did fear and a sense of dread overshadow me? I tried to shake it off and to praise, to be thankful, and joyful. But the heaviness persisted. I realized that part of it was due to the necessity of leaving dear Jerusalem behind, after having had so much difficulty in getting there. Our Pentecostal Conference was only four days away—and we had been within a short walk of Zion's Hill! Now, fast as the plane could travel, we were leaving it behind. Our first Tour Plan had called for a stop in Egypt enroute to Jordan. This, of course, was the reasonable route to travel. But, due to the large number on the tour, we had been shifted around and were obliged to backtrack and fly all those extra miles. However, I realize now that had this not been the case, I could not have tasted—in a small way—the sorrow Abraham, Joseph, Jeremiah, and others, including Mary and Joseph, must have felt in having to leave the Holy Land, their native land, and go "way down to Egypt's land." For Egypt, to the Israelite, was "the underworld," the "land of the dead." It was a place of darkness, sin and idolatry. It symbolized Sheol, and they both hated and feared it. And always, as surely as one went "*up* to Jerusalem," one traveled "*down* to Egypt." Indeed, throughout all Scripture, Egypt symbolizes "the world"—the state of being dead in trespasses and sin. No nation on earth, then or now, was so preoccupied with mortality and death, and so little with this present life. All its memorials and culture center in the ever-present reality of death. For the living, little is done. Its people are among the most underprivileged on earth. No nation built such monuments to death nor hoped more certainly for resurrection and an afterlife—yet all in vain!

None of us enjoy our "little journeys into death"—even though we know we are following our Lord as we make them. And few of us can approach them without a certain sense of fear. The enemy is quick to take advantage of this, so he began, on this particular day of which I write, to torment me that my appointment with the Lord in Egypt was an appointment with death. He revived the suffering and uncertainty I had undergone when the tour had been canceled before we left home. He tried hard to convince me that I would not return to Jerusalem, I would not be present at Pentecost after all; and indeed, I would never leave Egypt alive. He tried to make me think that the whole trip was just a sort of "mockery"—and that my little life on earth was going to end in most inglorious defeat in Egypt.

We grow accustomed to meeting our enemy on our "home base." But on strange territory we are at a great disadvantage. Flying through the air in a plane is still strange territory to me, and I cannot quite forget that the enemy is "the prince of the power of the air." So, dear ones, I am sure you can readily understand that I was undergoing heavy attack about this time, and knew that I simply must overcome *quickly*—if at all. Meanwhile, the plane had made a good takeoff in spite of me and my struggles, and as soon as it gained its proper altitude, and our seatbelts could be unfastened, our stewardesses disappeared, to begin preparations for lunch. I settled back, attempting to relax, praise and exercise faith to overcome these piercing thoughts. I waited eagerly for the cheerful voice of the pilot to give us the usual "welcome aboard" greeting. (This surely has a reassuring effect upon the jittery traveler.) It is customary for him to tell us about the route we shall be taking, the altitude of our flight pattern, and our approximate time of arrival at our destination. A weather report is sometimes added. I had heard that our flight would likely take three hours. Since we would cross a time zone, we would gain one hour on the clock. Hence, I reasoned, our arrival in Egypt ought to give us ample time to get settled in our hotel and have a good rest before dinner. However, the minutes passed, and no warm voice greeted us. Neither did we see any signs of lunch. Some grew very restless and began making processions up and down the aisles. Since the clouds obscured all sight of the land over which we were passing, all we could do was guess about where we were flying. One thing was certain—we were not over or even near Israel!

Many of the passengers, including Dwight, had let down their seats by now and were apparently sound asleep, or were trying to be. I never got over marveling at how casually most of these Americans took to flying. It was apparently as natural to them as walking. But not to me! No matter how I tried, I could not go to sleep. I found all sorts of new fears besetting me. Every odd noise in the engine increased my uncertainty. And when we had a few air currents and bounced about, I was really alarmed. What kind of a pilot did we have anyway? Our stewardesses spoke only broken English, so probably he didn't speak it at all. Was he well-trained? Was the plane properly serviced? Then I realized that for the moment I had forgotten all about the Lord being our pilot! Before long I was in a real travail of fear, and I had to fight hard to sit still and give no sign of it. About this time the clouds broke a little, and those who were awake began talking about the mountains below. I caught only a tiny glimpse, since the wing was in the way, but one look was enough! They were jagged and most ominous. A forced landing here would be impossible! A few began to get sick, and I knew that I ought to be very thankful that I do not get airsick. I tried to praise and give thanks and think about how very long and hard the journey to Egypt had been for sweet Mary, Joseph and Baby Jesus! I confessed my fears again to the Lord and it came to me that if He can give *"dying* grace," He can surely give *"flying* grace" too—so I kept thanking Him for such. If I am to fly, then let me fly in *faith*, in a God-honoring way, I prayed. I realized that I could do absolutely nothing about piloting this plane, but my *faith* might help in time of need, where my *fears* could not, and might even be destructive. It was an intense inward tussle. But gradually I grew calm.

Suddenly, the cabin door opened and the delectable smell of coffee permeated the air. Food was at hand! Of course waiting to be served is a little trying, but at least we knew it was coming. I was very weak and hungry, but so nervous that I wondered how I could eat. However, one does learn to eat aloft, regardless of how one feels about flying. The sleeping ones woke up when the trays reached them, and an air of good cheer settled about us. (We surely do depend a lot on mere food, don't we?) After all the trays were served, one of the stewardesses answered questions about our flight. Yes, it took three hours. Yes, we would gain one. We were flying via the mountains of Moab, the Sinai Peninsula and then across the Gulf of Aqaba and the Red Sea. She would point them out to us, and also the Suez Canal. We all felt better, now that we were located. And her conversation with us confirmed my suspicions that the pilot couldn't speak English.

When we did reach the Red Sea, almost everyone got up and leaned over those on the other side of the plane—to get a good view. I still couldn't overcome an awful feeling that a sudden shifting of weight might turn us completely over. But, apparently, this doesn't happen on passenger planes. I refused to add my weight nevertheless, lest I be the straw that proved too much. So I saw almost nothing but a blur of blue water. The Suez Canal was some distance to the north, and I felt I could get along without seeing it too—so I remained seated and tried to stay in prayer and praise. I sensed that I needed a real preparation in the Lord for what I would experience in Egypt. Our plane by now was flying over the long, lonely stretches of sand, somewhere near where the Children of Israel had once made their escape. But all this seemed unreal to me, up there far above it all. I was still in a spiritual exercise, calling upon the Lord for strength and grace and praising Him at the same time. I had no feeling of His presence, but I know He did draw near and give that which was needful for the experiences that lay just ahead. After all, He has proved Himself God in Egypt as well as God in Jordan and Israel many times in Bible history, and He's just the same today!

In time we began our approach to Cairo, and our stewardesses assured us that the weather was fine and we would soon prepare for landing. All the clouds were gone by now—for Egypt, as you know, rarely has any. Below left there was much golden sand and then, suddenly—most dramatically—it was bright green everywhere! This could mean only one thing: we were over the Nile area. Centuries ago Egypt was delineated as the entire tract of country which the Nile overspread and irrigated. The Egyptians are the people who drink the waters of that river—the longest river in the world. (It is over 4,000 miles in length!) A modern writer, Robin Feddan, has skillfully described the land as follows:

"Through a yellow-gray inclement desert, treeless and unwatered by rain, the Nile River has carved out for itself a passage. In so doing it has created the

217

landscape of Egypt. Standing on the desert hills"—(or seeing it clearly from a plane as we did, when we departed three days later)—"you look down to a belt of gentle green, narrowing or widening with the width of the valley and faithfully pursuing the course of the river. Below Cairo it expands into the fan-shaped delta"—(two-thirds of all the arable land is in this delta.)—"Egypt is like a layer of green wedged between an eternity of sand. *The sensation you receive of beholding a country at a single glance, of being able to stretch out your eyes and hands over an entire nation, is nowhere to be paralleled on earth.*

"In the midst moves the river—sometimes mauve-brown, again Nile green—dashed always somewhere with two or three white sails catching the sunlight. Alluvial and patient, the fields on either side receive its waters. From the mainstream the canals diverge like arteries and go about their necessary duties. From these again spread the lesser veins, and at last, gleaming under the sun, trickle the separate rivulets that the waterwheels raise to souse a single field or a few square yards of land. All this water moves by devious ways through an indescribable pathway of green, doubly lush by contrast to the adjacent desert, and variegated with yellow maize stalks, or the rich purple-brown of fields where a crop has not yet risen. From the desert, scattered forests of date palms, or clumps of sycamores, seem like the precise and miniature decorations on a geographical model or sand-table. Little mud roads, raised above the level of the fields, run determinedly up and down the valley, and, dropped in the belt of green at proper intervals, are gray mud villages."

On Sunday School days we had made such sand-table models of Egypt and other Bible lands—yet little did I realize how from the sky one can clearly see the Lord's own sand-table of Egypt! About this time we began to circle, much like the great falcon that is Egypt's ancient symbol. I cannot quite explain what happened to me then, but it was as though I entered another world, for the very atmosphere of Egypt was someway projected to me, even while we were in midair. I felt that time had been turned back thousands of years, and that I was going "home" in *time*. I was returning to a civilization that had truly played a great part in our own. I found this not unpleasant at all, as I had expected, but most entrancing and evocative! Mingled in it were nostalgic thoughts of Joseph, Moses, the Israelites, Abraham, dear Jeremiah, who died there in exile, and of Jesus. Pharaohs, slaves, pyramids and the

Nile all someway blended in this mood-tapestry—this magic carpet which carried me into the past.

Our "falcon" made a smooth landing and then a complete taxi-circle around the large, modern Cairo Airport. Flags were flying, as though to welcome us. And a large grandstand had been erected, decorated in engaging Egyptian style. I pretended, of course, that this was a sign to us. And I felt a strange sense of majesty, as though the Lord were "reviewing" us. The stewardess was more matter-of-fact about it and said it was for Sukarno, who was visiting at this very time. Well, we know him too, and Indonesia is one of "our" nations, so I felt very pleased about it all. We noticed the heat as soon as we touched ground, and it had steadily increased while we waited for the doors to open—it was stifling in that crowded plane. We were among the last to leave, because exiting too had become a "push and pull" exercise. I was still curious about the pilot. We surely must have one, and etiquette required that he at least tell us goodbye. I still hoped for a glimpse of that Arab through whom Jesus had been piloting us. What a surprise I had! "That Arab who couldn't speak English" turned out to be a strapping six-footer from Texas! There he stood, ready to give us a real Texas handshake and smile, drawling out, "It was nice to have you aboard, have a good time in Egypt!" He had been flying over there for several years, and was very anxious to come home, he assured us. Why he didn't deign to speak to us on the plane, I'll never know!

Although the field had few planes on it at that hour, they had parked us about three blocks from the entrance to the Passport office. And walking three blocks in that hot Egyptian sun, carrying all our hand luggage, was exhausting. Some even tried to run ahead on this too—to be first. I was quite content to be last, and almost was. Actually, there was no need for haste, for the first section of the tour, which had arrived more than an hour ahead of us, was still standing and waiting inside the entrance. And there was no air-conditioning! Neither were there any drinking fountains nor cold drink stands. And the only welcoming committee of which I was aware was composed of a bevy of large flies that rushed upon us with enthusiasm and loud buzzing. I remembered then that Egypt is famous for flies—so what could be more fitting or typical! One writer has said, "In Egypt only the *fly* is constant, impervious to time; Lord of house and land, it swarms unreprimanded on the garbage and the children's eyes alike. It is everywhere prolifically present in the thick air, heavy with dust." I can ensure you that this "welcoming delegation" was most unwelcome to us, but we were helpless to ward it off, since our arms were still filled with coats, purses, cameras, flight bags etc. I couldn't help but wonder how many diseases such little "beasties" packed around with them.

In the midst of this jam of tourists, a representative from Compass Tours was trying hard to quiet us all down and talk to us. And after the chorus of greetings died down, we were assured by him that everything possible had been done to make our visit in Egypt most enjoyable. However, due to the large number arriving in one day—over 600 in our tour alone—the hotels and facilities were overcrowded. Also the customs officials were overworked. So we must all cooperate and follow instructions in order to get through the lines with as little delay as possible. This cheered us up considerably. The lineup was made and we began to pass by for inspection of passports, the Tour Director assisting in identification or answering questions. There was no place to put our luggage down and leave it with safety, so we had to stand in line clutching it as best we could. Eventually we put most of ours on the floor and kicked it along as we moved forward at a snail's pace. How very wearing this long lineup proved to be—for it was hot and close in that waiting room. As usual, the grumbling began with a murmur and gradually arose to a din. My heart went out to those who were ill or incapacitated in any way, and to the older, weaker members of our tour. Short of fainting or going into a coma, there is no way to escape these grueling ordeals, even on first-class tours. In fact, ours was hastened along, I am sure, due to the large number of us. But even then it took time. When we finally got through this lineup of officials, all of whom eyed us suspiciously and with apparent displeasure (no doubt due to being overworked that day) we were told to surrender our passports to the Tour Director. This troubled us not a little, for right on the passport, we are instructed and warned that never, under any circumstances, are we to allow our passport to be out of our possession. Yet, we also had to obey the Tour people—so we dutifully complied, but not without misgivings.

We then passed through another gate—but not to freedom! To our dismay, the next process involved another lineup. We must fill out long printed forms. By this time I was beginning to get disgusted with Egypt—for they had required two visas of each of us, instead of the usual one—and now these long forms! We were asked an amazing variety of personal questions, including how much money we had with us and how it was being carried—in currency, travelers checks etc. We had to list it to the penny. Also our jewelry, cameras and other personal possessions, and their value were listed. (I guess this was so we could not falsely claim anything stolen.) Since these officials spoke only broken English, and we did not know enough Arabic to ask questions, everyone got very mixed up and frustrated about all this questioning. After filling these out we had to pass by and affirm that our statements were true. Again we were treated with suspicion. But about this time I recalled that I had read that the United States is the most difficult of all the free countries to enter, and treats tourists the worst—in the matter of forms, visas and examinations. So I decided not to be upset at Egypt after all.

Relieved to be free at last, we hurried along to a large outer room. But again we were stopped. Now, it seemed, we must all stand about and wait until our large luggage arrived. No one could go until his bags were identified and possibly examined. This was just one thing too much to bear! Tempers flared and some of our Tour behaved most unChristlike. By now the flies, heat and wearisome waitings seemed intolerable, and the precious time, which we had hoped would be used in getting located, had all been spent at the airport. Fortunately, there were so many bags, they did not open any, though I observed them thoroughly searching those of a few tourists traveling alone. I praised God that we found our bags quickly and could then go through a turnstile into the blessed outdoors. I hoped getting out of Egypt wouldn't be as hard as getting in! It was still hot and most depressing in the heavy air. But at least it was fresh. Again there was more waiting to do. Now we must stand by the buses, and could not get on them, until our luggage was brought out and placed by the bus on which we would ride. Oh to sit down a while! Something happened, it seemed, to one of Dwight's bags and it did not appear. So he decided to go back in search. He got back in all right, but when it came time to go out, they refused to let him out. He had several very difficult moments, but finally the Tour Director came to his rescue. By this time even Dwight was shaken up and running out of patience. It took all the grace I could muster and a rigid self-discipline to wait

221

patiently until we finally found our bags and were told which bus to get on, and allowed to sit down and wait there until we were all assembled.

The ride to Cairo was delightful, and the cooling breezes revived us. I could readily see that it was a large and very exciting city. And there were many beautiful flowers and trees along the way. The architecture, Moslem in character, mingling with modern buildings, gave it a real charm. I could understand why it is the capital and center of a vast Arab world—a city of more than 500 mosques! I realized that now I was very near to the heart of the Moslem peoples, and it both thrilled and awed me a little. It was harder to *feel* "this people is my people" than it had been when I had visited them in the Spirit. But by faith, I knew it was true. Our party had to be divided, we were told, so about two dozen of us were dropped at Hotel Everest, which was in the very center of the city. We followed an Arab guide to an inside corridor of a large building, and were placed, a few at a time, in a small elevator, operated by a widely grinning Arab boy, who broke into laughter at my "*Naharak Saeed*"—which I hoped meant "good day!" (In Jordan they had told me to use this in the afternoon.) Then, suddenly, we were going up, up, up, with breathtaking and stomach-upheaving rapidity! Elevators have a special place in my category of childhood fears, since I used to dream often of being in one that dropped endlessly. I was surely glad to get out of that one in Cairo; but dismayed to find that we were parked on the 15th floor of an old building! And me, with my sense of acrophobia! No wonder they called it Everest! I couldn't help but wonder if they used steel to reinforce buildings in Cairo, and if they ever had earthquakes!

The lobby seemed very small to me, as we all crowded in it, but soon we were following a bellboy down a narrow, rather musty smelling and dimly lighted hall to our assigned room. At first glance it appeared small, but fairly modern. And it had large French doors and a balcony—oh wonderful! Here was another balcony on which to overlook a city. Since childhood I have doted on balconies, and almost never have had the pleasure of enjoying them, so I put my things down quickly and hurried out. Two balconies in a row seemed too good to believe! I praised the Lord with much joy and then cried out in amazement, "Dwight, come and look, can those possibly be the pyramids over there?" In the lovely glow of late afternoon our eyes beheld an enchanting scene on the horizon—the misty blue Nile river and the ancient pyramids! I felt I must be seeing a mirage! But he assured me it was real. My heart overflowed with praise to the Lord for giving us a room that afforded such an exciting first view of these wonders so dear to my heart. A famous writer has written, "The finest way to see the pyramids for the first time is at a distance, suddenly." And I agree. The impression this view made is one of the most beautiful of the entire trip, perhaps second only to that first view of the Mount of Olives by night, and the view of Zion's Hill shortly after dawn—a view four days yet ahead! The coloring of the landscape was spectacular! Many authors have tried to describe it adequately, and have despaired at doing so. It is unbelievable, almost fantastic the way the varying shades appear. One wrote: "The fall of dusk upon the Egyptian scene is an event of unearthly beauty. Red embers, coral, rosy flush, rainbow—rose-pink changed to green and gold-grayish opalescence and . . . gradually . . . to blue velvet." I couldn't tarry on this first afternoon to watch all this procession of beauty, but I prolonged this rare interval as long as I dared.

Far below me the unbelievably complex and noisy traffic of Cairo's crowded streets was at rush-hour peak. What a strange cacophony to accompany a scene as ancient, majestic, beautiful and inspiring as could be found anywhere in the world! It is believed that Abraham's eyes looked upon those pyramids—at least on one of them, and Joseph's eyes beheld them. Moses most assuredly knew them well. And all these had drunk the sweet salubrious waters of the Nile, including Jesus. It is said that if one drinks of its waters, no other water will ever again satisfy him. As I stood transported both by the natural scene and the moving of the Holy Spirit within, I knew to the depths of my being that I loved this land too, and that the Nile would forever be dear to my heart. I was "way down in Egypt's land." And I was also standing in a high place, as though on a mountain, overlooking another "promised land"—a land for which we and countless Christians have prayed, a land which eventually will know and praise the Living God! I did not realize then, as I do now, that it was in this very land and by this very river that the Sacred Name—YHVH—had been first proclaimed among men, and that here its power had been first displayed. No wonder I felt awed and almost rapt with wonder!

NIGHT ON THE NILE

Among the beauties and wonders of nature, nothing attracts and satisfies me more than close, personal contact with a swiftly flowing river. Therefore, it had been exciting, in looking forward to this trip, to anticipate visiting some of the most famous and beloved rivers of the world: the Thames, Tiber, Seine, Jordan—and, most certainly, the mighty Nile. However, as you know, my brief introduction to the Jordan had proved most unreal and, in a sense, disappointing. So I did not expect to find the Nile very inspiring either, since Nile-green is not a color I much admire, and I had also read how muddy it became at flood season. However, after having had this first breathtaking view of it from my 15th floor "Mt. Everest" in Cairo, I found awakening within me an intense desire to see it at close range. I knew that this misty swath of blue-gray chiffon draped before the pyramids was one of the world's greatest and most fabulous of rivers. And even from a distance its waters were drawing me with a surprising magnetism.

I am sure that this love I feel for rivers is not just because I have come from Southern California, which is famous for its unusual rivers. Even between Idyllwild and Los Angeles there are several of these to cross (i.e. San Gabriel, Rio Hondo, Los Angeles.) Of course the thing that makes these rivers so outstanding and astounding to tourists is their *novelty*—they are completely dry! It is said that nowhere else in the world would the Chamber of Commerce dare to build bridges and put up signs about rivers over dry gravel and sand beds. But we know that this is not just advertising, they are actually on maps too. And this is because on about four or five days out of the year our rivers *do* have lots of water in them—muddy, churning, debris-carrying runoffs from rain in the foothills and mountains. But there isn't much about these floods to inspire one, hence we each must seek our river-joys in other areas.

Yet, I am sure that my love for rivers goes much deeper than the lack of them in our land—it is rooted in a deep, primitive association of rivers with life itself, and with beauty, power and blessing. All children are attracted to streams and rivers, and I was no exception. My mother well remembered that from the time I was two years old I repeatedly ran away. And I always headed for Cherry Creek, a river that flowed through Denver. There, on several occasions, I was found, leaning precariously over the bridge railing, apparently fascinated by the lively waters beneath. What more perfect symbol of life can be found than a lively, dancing, singing river? Oh the scintillating sight of it—the music of its many sounds! It blesses and gives light to all it touches. It has a magic way of carving out a path of beauty—and sometimes of magnificence beyond all description wherever it goes. Who but could marvel, as have I, that the Merced River (Our Lady of Mercy) could spend eons of time painstakingly cutting and polishing such a jewel of beauty and splendor as Yosemite Valley? Or think of the King's River, flowing through the Himalayan-like majesty of the High Sierras, and producing such a display as King's Canyon. (Now there's a river after my own heart! It moves me as none other—so high, so fast, so filled with dash and spray!) Or share with me the wonder of the narrow, but very purposeful Arkansas River, as it forces its way through the great Rockies in Colorado and creates a place where all may stand with wonder—the Royal Gorge! Have you followed, as I have, the great Columbia River (The Dove) in Oregon, and almost swooned at the lavish display of falls, rills and flowers along its banks? Utah has its Virgin River, the heart of Zion National Park. (It's a small stream, but oh! oh!) And of course there is the greatest and most splendid of them all, the Colorado, beginning like an icy ribbon in the Rockies and pursuing its course with such violence, determination and utter disregard of man that engineers have spent a century trying to master it. All along its way it has sculptured such magnificence as one views throughout Utah and culminates in its last prodigious georama—in the incomparable glory of the Grand Canyon. Were this a treatise on rivers, I could continue on and on: the Niagara, and its rapturous falls; the Amazon and glorious Angel Falls; the mighty Mississippi; the St. Lawrence and countless others. Yes, rivers are doubtless the most perfect symbol of the Holy Spirit's Life and Power human words can find, for our Lord Himself referred to the Spirit's coming by saying, "Out of your innermost being shall flow rivers of living water."

It is said that if men desire to build a great city, they must be sure to locate it by a river. And I heartily agree. New York has its Hudson; Washington its Potomac;

London, the Thames; Paris, the Seine; Rome, the Tiber and Cairo its Nile. And God Himself has built His eternal city on a river, "There is a river, the streams whereof make glad the city of God, the Holy Place of the tabernacles of the Most High." This, for us, is the true River of Life. And Isaiah and Ezekiel and other prophets have declared the beauty and power of this river. No wonder we love rivers and thrill to them—it is quite scriptural, I assure you!

But return now from this breathtaking tour of rivers and stand again with me on my balcony on that memorable day in Egypt. Still gazing at the scene, I whispered to myself, "*Bukra* (tomorrow) I shall see the Nile and the pyramids." Then I recalled having overheard the Arab guide say something about taking a moonlight boat ride on the Nile this very night! How thrilling! I turned to ask Dwight about it, but he was interested in running water of another nature, preferably hot and at hand! He had already unpacked his bags and was preparing to shave and shower. Since no electric outlet was available, he was ready to lather up with a safety. The problem occupying him was that though we had what appeared to be an Egyptian variety of washbasin, with the customary two faucets, both were trickling feebly with only one variety of water—lukewarm.

"Something's wrong, Sweetie," Dwight was saying, "it just doesn't get hot." I told him that likely everyone was washing or bathing at once, and I rejoiced that our room afforded such luxuries as a private washbowl and shower. The novel design of our suite may interest you, as it did me. Upon entering from the hall, we were at once in our bathroom. It had a cement floor and a large drainage hole in the center. On one side was a mirror and basin. On the other, behind a narrow curtain, a shower was placed in the ceiling. The balance of what one usually expects in bathrooms was obviously located somewhere else and, presumably, would not be private. But it was nice to think that we could bathe, wash and do our laundry here. That is, of course, if the water finally got hot, which it didn't. Another novelty was the total absence of towels. Stepping up from this bathroom we were in our very tiny bedroom, having twin beds and a mirror and shelf—no drawers or stands of any kind. There was also a built-in closet. And that was it.

About this time, without knocking, in walked a widely grinning teenage Arab boy. He was carrying our towels, and he greeted us with the usual "halo." Then he stood grinning even more broadly, his eyes beaming with pleasure. It was apparent that he expected us to be overjoyed at his presence. So we tried to be. But after a few efforts at conversing, we discovered that his vocabulary was limited to "yes, no, halo, please and thank you." I was in no mood at that moment to try out any Arabic and have an impromptu lesson, so I kept smiling and nodding. But he had no intention of leaving us, it seemed, for he just stood and looked over all our belongings and us with frank and unabashed curiosity and interest. Dwight had to sort of usher him out by pointing to the shower and indicating that he was about to undress and bathe. So, reluctantly, he left us. We knew he wanted a tip, but it is best to give them when leaving, so we had to disappoint him.

"You won't believe it, you just won't believe it," Dwight called out to me. "Put on my glasses and come here!" Curious, I hurriedly found them and stepped into our "bawth." He was right! I didn't then and I still don't now! He was holding two of the grimiest towels I have ever seen. All I could think of were my mother's old dirty mop rags. And she wouldn't have reused them then without bleaching. "They must have been washed in the Nile without being pounded or rinsed," I observed. I hated even to touch them, but I did and saw that they were stiff as well. For a moment I felt really sick. Then I remembered that we had each brought our own towels for just such an emergency. So I reminded Dwight of this and went to get his. We simply could not use these! But Dwight was now aroused to investigate our room and bed. The light was dim, but the rugs were obviously very soiled; the spreads, likewise. Ugh! He warily turned his back and found a thin, dingy cotton blanket, and a pillowcase of the same shade as the towels—evidently the Egyptian variety of "tattle tale" gray-charcoal! The sheets, though, really took the prize. They had been torn out of coarse yellow, unbleached muslin, and at least did not look very soiled. They were not hemmed at all, and were barely wide enough to cover the mattress, which seemed to be stuffed with straw. There were no pads under the sheets, and no springs under the mattress—not even the flat kind they had in Jordan. We just stood and stared for a moment. We hadn't been able to carry along our own sheets and pillowcases too! So I decided to use a heavy slip for a case, and resign myself to using the sheets. I realized that missionaries all over the world have had to endure conditions worse than these. So why should I be so squeamish? Dwight, by now, was putting his shirt back on and heading out the door. "It's time for me to pay a little call on the manager," he said, "and find out about a few things, including the hot water." "Get our mail," I called after him. I needed a "transfusion" about then.

I flew into my unpacking, sorting out my laundry, trying to praise and be cheerful. But I suddenly felt terribly flat and exhausted. And thirsty! This was getting to be an overwhelming problem—the lack of drinking water. I should have realized that one can purchase bottled mineral water in most hotels. Some tours supply such to their members. (However, my neighbor, who has taken luxury tours, says that they always wondered if the maids didn't fill the bottles at the tap anyway.) I needed my vitamins and a drink about now, and was loathe to drink out of those ancient looking taps. After a struggle, I decided to risk it. The Nile, in itself, is wonderful water, they say. But it does get pretty dirty. Surely Cairo must have a system of filtering it, I reasoned. So I drew some water and took my vitamins and gulped it down. Right away I felt a little sick. Fear, no doubt! I must eat and drink in faith, nothing doubting, I told myself. I recalled the promise about nothing deadly hurting us. Someway it seemed hard to believe about now.

Once unpacked I hung my clothesline on the balcony and rinsed out a few things in cold water. Then I spread out my coat and lay down on it, feet up on the wall. Ah, blissful! Even ten minutes' rest was priceless! This was about all I had. Dwight returned really agitated. The hotel was in an uproar, it seemed. He reported that at least a half dozen couples had refused to accept their rooms, after seeing how dirty they were. They were now in the lobby, surrounded by their bags, protesting to the clerk. He had told them that this was a *third* class hotel—so what did they expect? They insisted that our Tour contract called for first or tourist class hotels. Hence, our rights had been violated. They were demanding that the Tour representative be contacted and that he move them into a suitable hotel. So far they had gotten nowhere. They wanted all of us to take a similar action. And Dwight wondered if we should not join them. Ah me! The thought of repacking the bags, I guess, was just too much. I wanted to cry. And there was my laundry dripping on the balcony! I am glad it was, for it held me steady. I reminded him that we had known ahead of time that some hotels would be bad. He replied, "But the worst part is the food—it won't be safe here. You should see the dirty boys that are the waiters. What are the cooks like?" This did affect me and I was shaken. Then I said, "Are the McKeechems among the others?" I had been drawn to this couple from the first—and they were always in a sweet, positive attitude toward the Lord and our companions. I knew they were in this same hotel. "I didn't see them," he said. He too lay down and for a few minutes we were quiet. Then he said, "Well, let's clean up the best we can and go out to the lobby. They are ready to serve dinner, if we decide to stay. And, by the way, there isn't any hot water available, except between 6.00 and 8.00 A.M. And there is no mail," he added. This *was* a disappointment.

Once in the lobby, I could feel myself being swept along in a wave of turbulence. Our Tour companions were putting on quite a "floor show," and they had an interested audience composed of hotel employees and other guests. One woman in particular was "holding forth." I could just picture what she would be like on a church board. So far she had been a leading "committeewoman" all along our journey, heading the chorus of complaints and suggestions for rectifying conditions. I felt sick at heart. Then I saw the McKeechems approaching, and the picture is still vivid in my mind. Wearing her usual bland little smile, and walking daintily on high heels, Christine looked as fresh as a flower. Her tall handsome minister-husband was right behind. Without once pausing, they made their way without haste through the agitators and others, and walked deliberately into the dining room and sat down. As though by mutual consent, Dwight and I fell in right behind them. Two other young ministers, to whom we also were drawn, followed us. The agitators gasped! One of them said something about our being crazy, and that we would get poisoned eating in there. But we went on in out of earshot.

About six youthful waiters rushed to our side. Obviously they were relieved that somebody was going to risk eating with them. We were showered with attention. They did look dirty, and I prayed for special grace. Meanwhile we all sat smiling and conversing as though nothing at all was wrong. I told Christine that we had been waiting to see what they would do, since we had found them so well-behaved on every occasion so far. She said, most humbly, "Well, we're just going to enjoy every minute of this trip, no matter what happens, we're so thankful to get to come." Then she told us that their congregation had made great sacrifices to send them on this tour, and they did not intend to spoil it in any way. I mentioned that this was also true in my own case, and that we wanted to praise and glorify the Lord each step of the way. The others agreed, and we soon were laughing and having a fine time. And by this time a number of the colored tour members were seated at another table. The first course was soup. Since it was boiled, I felt it was safe. And it was fairly good. Then came salad. We all had been warned not to eat any raw vegetables or fruit, so we only played with it. But how very nice it looked! The meat we risked too—not being sure what it was—but we ate it anyway, and a cooked vegetable. And I enjoyed the bread, which was in little rolls. By the time dessert was served I had forgotten that we were not supposed to eat any milk or milk products while in a doubtful area. It was some sort of sherbet and little cakes. I ate most of it and drank French coffee. Not bad! (In fact, I have had far worse meals in L.A. and paid a good price for them.)

Even before we had time to finish eating, our Arab Guide was at the door, telling us it was time to go boat riding. The agitators were gone by now, but their bags were all still sitting in the lobby. They had gone to the Nile Hilton for dinner, we learned, at their own expense. We hurried to our rooms and again checked our bags to be sure they were locked. And we looked to the Lord to protect the things we had unpacked, for we still felt odd about our surroundings. Then we joined the others and were put into a taxi for a short ride to the river. "This is all free," our guide said. But we knew he meant only the taxi. The "bite" would come at the river. Caught in the swirl of Cairo traffic, I found it something else I just couldn't believe. It is simply indescribable! For one thing, everyone that has a horn sounds it practically continuously—and this is no exaggeration. The din is like New Year's Eve in Times Square. The streets were rather wide, but crowded with a great variety of vehicles, ranging from ancient—drawn by donkeys and camels—to modern. The pedestrians seemed to be waging a constant resistance strike against all drivers of vehicles. They deliberately get right in front of them and even make faces at them sometimes—not dodging out of the way until the last possible moment. The drivers lean on their horns and go all the faster, by way of retaliation. I was simply so frightened for those on foot that I forgot all about being afraid for us. The drivers always missed—but meanwhile we kept gasping at the near-rundown of at least two dozen pedestrians of various sizes!

Fortunately the ride was short, and when we stepped out of the taxi, the view was exceptionally expansive. All along the Nile beautiful hotels, apartments and other buildings have been erected. And a parkway of trees and flowering bushes has been planted on its banks. A large well-lighted bridge connects its banks with Gezira Island, located about in the middle of the river. And on this island a really magnificent tower has been erected. We had seen it from the hotel and noted the large revolving torch at the top. Cairo is very proud of this tower, and it attracts the admiration of visitors from all over the world. At night it looks especially thrilling, for it is well illuminated, and its brilliant revolving torch, lifted high over the city, affected me as only torches can. (So often we have seen God's light symbolized by a torch.) But this tower and torch reminded me of the tower of Babel and of a vision I had once had about Egypt and that tower. I know that originally it stood in Babylon. Yet, in my vision, I had seen it in the midst of Egypt's sands! Cairo is indeed the center of the Arab commercial and political world, and the stronghold of the Moslem religion. In any case, this torch seemed to me to be the light of antiquity shining out to the world, teaching all men, through the experiences of Egypt's past and present, the hopelessness of the nation that does not walk in the light of YHVH: if ever a nation—from the king to the lowest servant—had opportunity to know the Living God, it was Egypt! God sent both Joseph and Moses, as well as many other witnesses, into that land. And to that nation—then the greatest on earth—both His name and power were declared. Later, in the days of the Apostles, they too visited Egypt and declared the Gospel unto the people. Yet today Egypt is a benighted Moslem nation—with about 90% of the people blinded and bound by its traditions!

The beauty of the Gamaa Bridge was fascinating to me! It is brilliantly illuminated and the arches beneath it glow also, giving it a fairytale splendor. The Nile itself picks up all these lights of the bridge and the esplanade and reflects them with multiple refractions. The effect upon the one who first views this dazzling sight is almost hypnotic—at least this was my experience. I had seen a Nile Hilton ad in a magazine, showing this area, and just couldn't believe Cairo had such charm. But it is true! I longed to stand and worship and revel in this Arabian-Night-on-the-Nile fantasy. I wanted to relate it to the Lord and to His Kingdom splendor, by faith, even though by fact it does not seem to honor Him. I wanted to lift up His name there by the Nile, as Moses and others did so long ago. But, of course, our guide was impatient to hustle us along toward the boat-ride. I am never very interested in embarking on a boat, so I lagged behind, delighting in the trees and flowers, and in observing and trying to greet the people strolling by. I noticed that they all stared at us with either interest or curiosity, but when we smiled at them or spoke, they smiled and then quickly looked away. They were dressed in both native and western-type clothing, but all had dark faces. One very pretty girl gave me an especially penetrating glance, then turned, after passing us, and looked so longingly in our direction that I called out to her in Arabic. She immediately left her companion and ran toward us eagerly, returning my salute in Arabic.

"You are an American?" she asked, speaking now in English, "and you can speak our language?" "Yes, I am an American, but I know only a very little Arabic," I told her. "However, I am studying it and trying to learn it." This pleased her very much and she began to explain to me what a wonderful language Arabic is, and how regrettable it is that Americans do not seem to want to learn it. She urged me to continue with my study of it until I could read their classics. (This really made me gasp—I have such a time even reading their alphabet!) I felt a strong flow of love from my heart to hers, and wanted to hug her. She seemed to feel the same way toward me. She told me that she was from Syria and was studying at the University of Cairo. And she taught me to say, "*En Neel gameel*," I believe this means, "The Nile is beautiful, wonderful, etc." We said it together again and again. Then she ceremoniously waved her arms in a gesture of invitation, and said dramatically, "Welcome to Cairo and welcome to the beautiful Nile, the river of peace and love."

As she spoke these words, I felt a witness in my heart, as though the Lord Himself were extending His love to me in this strange land. Here was a charming girl from Syria—another country we love—acting as His messenger. Great joy filled my heart, and we embraced. By now, her companion, an Egyptian girl, had joined us and was also smiling and nodding to me. The Syrian girl continued to talk about Cairo, and when I mentioned Egypt, she was quick to tell me that it is now called, "The United Arab Republic." I told her that I realized that its name had been changed, but that I have learned to love it by its name in our Bible. I explained that I was visiting there with a Christian group, and that we were especially interested in the land because Jesus Christ had once lived there, as well as many others who are important in our Scriptures. She was quick and proud to tell me that she was Moslem. (I noticed from my first contact with the Arabs this proud attitude. They are seemingly more open about testifying of their faith than we are. They also bring Allah into every conversation, no matter what it is about.)

I told her that Jesus had put a special love in my heart for the Moslem people, and that He loved them dearly. She seemed touched. She was eager to visit America too she assured me, and to know us better. I wanted to continue this conversation, but Dwight was tugging at me—all our companions had disappeared! So with reluctance I bade her, *"Ma'a es salame"*—a form of goodbye, and hurried on with him. But I can see her bright eyes even yet, as she stood looking after me, waving until we turned to go down to the boat landing.

I was fairly intoxicated with the love I felt toward her, toward this land and people—and even toward the great, dark river moving so quietly, yet so swiftly beside us. There were no rocks to cause splash or spray, so its song was low and quiet, almost ominously so, I noted, as we approached its edges. There were few lights here for it to reflect, and it was now not so beautiful. I was awed by its width—when I saw how far it was to the lights on the island—and began to doubt that I wanted to go sailing on it. Suddenly I felt that perhaps night on the Nile held peril as well as "love and peace." All too brief were the moments of joy! I was no longer intoxicated by the River of the Holy Spirit flowing through my heart; instead, I was aware of being intimidated by the dark River flowing at my feet!

The Nile is the longest river in the world (more than 4,000 miles) and one of the few that flows from south to north. Hence, in Egypt, one says, "Up south, and down north." This, of course, makes it seem upside down to us. It is difficult for me to picture the magnitude of such a river! But when I read that its length would be a *distance equal to that from Los Angeles to the shores of Eastern Greenland*—I was simply amazed! For thousands of years men have disputed about where the Nile really begins. In one sense it begins in Khartum (Sudan.) Here the Blue Nile, flowing clear and bright blue from the mountains of Ethiopia—or reddish brown in flood time—meets the grayish-green White Nile which comes from the lake region of Central Africa. The sources of the Blue Nile were easy to trace, but it is only in comparatively recent times that explorers have followed the White Nile to the ten tiny virginal springs that leap to life on a grassy hillside at about 6,700' above sea level in the Central African highlands of Ruanda Urindi. "There, in the swamps, among beautiful white, blue and crimson water lilies, it begins its descent and moves on past rapids and waterfalls from Albert Nyanza (lake), like a gigantic mill race, through the spectacular gorges of the 'Victoria Nile' which flows into Victoria Nyanza. From thence it moves on to the headwaters of this lake and then descends through the Sudan to the place of its wedding with the Blue Nile. For thousands of years the Nile has challenged explorers, historians, and scientists. I think it is not too extreme to say that the Nile has influenced, somehow, every person living in our Western World today. I was drawn by the lure of this great stream that blends the unfenced zoos of Africa's upland plains with the ancient cultures of Egypt and the Mediterranean, compounding history and natural history unmatched anywhere on this globe."

—John M. Goddard, scientist and explorer

BAPTIZED IN THE
RIVER OF EGYPT

Almost every picture we see of the Nile River depicts one of three characteristically Egyptian symbols: a palm tree, a pyramid or a felucca. Often all are included, along with a camel or two and some *fellahin* (as the farmers or common people are called.) We readily recognize the date-palm trees, for they are familiar to all Southern Californians. And most of us realize that they were originally introduced to our state from the Middle East—mainly from the Arabian lands. (Sometimes, as we have driven through Indio, I have tried to picture a large river flowing alongside the palm groves, and have thought how much like Egypt the landscape would then appear.) We can also quickly identify the pyramids as Egyptian for even though our dollar bills all display the Great Pyramid, every schoolchild knows that these famous stone tombs are found in Egypt. (It is true that there are Inca and Aztec pyramids also—but these are not familiar to us, nor do they compare in age and magnitude with those of Egypt.) The felucca, however, is not a familiar sight to us, even though it is as ancient and Egyptian as the pyramids. For thousands of years this particular type of boat has been sailing up and down the Nile. And it is still in use today, lending its special air of antiquity and charm to the river. Feluccas of various sizes and types were commonly used throughout the entire Mediterranean area in the bygone days. They are described as: "A vessel of Arab origin having lateen (triangular) sails and oars." The Egyptian felucca—which the Arabs spell *faluca*—is usually small and shaped somewhat like an enlarged rowboat. Often it has only one sail, though the larger varieties carry more. These tall white sails are shaped like an elongated pyramid, and blend most fittingly into the typical Egyptian scene. The faluca's grace in sailing is comparable to that of the palms growing along the shore. Its geometric pattern often allies it with the pyramids. And when seen in the dazzling sunshine of the Egyptian day it appears like a large white-winged bird silently skimming over the Nile-blue—or green—waters. A beautiful and evocative sight indeed!

241

However, when it is seen in the darkness of Egyptian night, it presents an entirely different aspect. And I can still recall how my first view of a faluca affected me. The little dock at which it was moored was poorly lighted. And the Nile was like a sea of ebony before us. Several other falucas, laden with members of our tour, had already set sail upon it. I could see them far out upon the river, ghostly in appearance, their tiny lanterns seemingly as small as fireflies. Even at close range the boat looked frighteningly small to me. (I had supposed that we would sail the Nile on some type of excursion boat, I guess.) High on its mast an ancient "one-candle power" oil lantern hung, and on the dock, with one barefoot on the boat, the very black Nubian boatman stood with hands outstretched to help the sisters aboard. In that eerie light, his face was shining and his eyes were luminous. He, like the boat, looked ghostly-clad in a *galabiyeh*, the long white nightshirtish sort of garment Egyptian men wear. I was swept with a wave of apprehension as Dwight edged me along toward the faluca. There was no time for thought, nor chance to draw back—though I wanted to turn and run.

As I relive this experience now, I can "stop the camera" and hold this moment as long as I want to, trying to recall and analyze what I felt. (This is one of the wonderful things about memory—its power to lengthen or shorten time!) I think that I must have connected this River and boat and its boatman with the legendary River Styx and the boat that carries one over the Stygian darkness of death; or with the accounts I had read of how the Egyptians placed the dead at times on a boat and transported them across the Nile to their "City of the Dead." The black Nubian was like a dark death Angel indeed! In any case, a strange sort of fear took hold of me, and I needed the grace of the Lord in addition to all the self-control I could muster to board it, sit down calmly and act normal.

There were about ten passengers altogether, and we seemed to weigh down the faluca considerably. It had sat low in the water even when unloaded. Now the river came almost to the edge of the boat, and there was no railing between us and it! We all sat on a little lower edge which ran the full length of the boat on each side. Our companions were quite gay and carefree, and apparently each thought this was a great adventure. Of course there were some complaints about the lack of moonlight the guide had promised. But I had not expected any. The Lord has long made us conscious of the phases of the moon, and this was only the second night of the new moon, so it had retired early, as is its custom at such a stage. There were many stars, I am sure, but some way they seemed very far away—as did the shore and the Lord.

My fears became like a travail, as the wind grew stronger and the faluca began to toss and dip about. Then, suddenly, I was in the grip of sheer panic—for I realized that perhaps it was not the will of the Lord for us to be on this boat ride at all! It was definitely not a part of the tour, but an added "pleasure." And we are not to be mere pleasure seekers. I had rushed into it without taking time to seek His will. Added now to my fear of the Nile was my fear of having displeased the Lord, placing our lives in jeopardy by a human whim. I realize that most of you would think it strange that a little ride on the Nile could be such an overwhelming trial to me. It candidly reveals what a big baby I am! Yet there was more than natural fear and danger around me this night—there was something supernatural and most tormenting in the atmosphere. The discovery that I had again made a move without being sure of the Lord's will unnerved me almost completely. I have long feared getting out of His will. I realize that great losses can be incurred by making even small steps in self-will, and that often our sufferings and penalties result from taking things into our own hands. But there was also a sense of awe of Him and of the gravity of displeasing the Lord which was most weighty—I am glad to know this fear of the Lord. However, I cannot honestly say which fear was predominant that night—the fear of the Lord or the fear of the Nile. I was just frightened clear through!

No one else on the boat seemed to feel a bit scared. They kept joking and conversing with the friendly Nubian, who could speak English fairly well and was eager to talk about himself and his people. He told us too just where on the river Moses was supposedly found—but all I understood was that it was a little distance to the north of Cairo. He made it plain that the Nubians are a proud and ancient people, and that the present Arabs who live in Egypt are a mixture of many people, and are not true Egyptians. It was apparent that there is race consciousness in Egypt too! Nubia, of course, is in upper—or southern—Egypt. It is a land that seldom if ever sees rain, but relies entirely on the Nile. I recalled that the Nubians were noted for their stature and physical beauty, and that they were more peaceable than the Egyptians, hence they were often captured by them in the days of the pharaohs, and made

243

slaves. There was something in the bearing and attitude of this Nubian that revealed to me some of their ancient and innate sense of dignity. I found myself being drawn to him by the Spirit—and I wondered briefly about his soul and his spiritual state.

The nearer we drew to the banks of the island, the safer and calmer I felt. There was a romantic-appearing outdoor cafe along the river's edge. And in the garden I could see various people sitting at little tables and walking among the flowers and trees. Candles, lanterns and little colored lights illuminated its beauty and Arabic music pervaded it. It seemed incredible that these people were so apparently safe, happy and at ease just a few hundred feet away, while I was tossing about on a ghostly faluca, wondering if I would escape alive from the threatening embrace of the Nile! How I wished we could land the boat and get out on dry land. But about that time the Nubian asked a rather overweight sister to move from one side of the boat and sit exactly in the middle. He did it very graciously, but she was obviously embarrassed. We soon understood why he made this request for to our amazement he got up on the very narrow edge of the boat, walked to the other end and began to shift the sail. This caused the faluca to dip way over on one side, and I gasped, feeling sure he would fall off. My fear was not entirely for him—whatever would we do adrift on the Nile without a boatman who knew the mystery of how to navigate a faluca! I had a glimpse of the horror it would be to try to pull him out of those dark waters without capsizing the boat. I could just see everyone rushing to one side, as they did on the plane. He sensed my concern and smiled at me reassuringly.

"I have been sailing falucas since I was a little boy," he said, "have no fear. I get it turned around fine!"

I looked at the others and realized that some of them were not feeling very confident about then either. But the boat righted itself and sure enough, we did turn around and start back for the other side. For a few minutes I relaxed then, and began to feel ashamed of my lack of faith. Even if I was out of the will of the Lord, I must remember that He is merciful and He will spare us. Thus I reasoned. But an odd sense of having lived through this experience before haunted me. My mind was flooded with tormenting thoughts. And a voice seemed to say, "You have come to Egypt to be taken into death—and this is the river of death." I was almost convinced by the deceiver that my journey was about to end. For I realized that my swimming ability could never cope with the deep, swift waters of the Nile, especially when I was fully clad. I could hardly breathe by now even above water! However, it was not until I began to write this account that I realized how truly I *was* reliving past experiences that night, though in a subconscious way. When the Spirit revived my memory, one of these experiences came back clearly: I was fifteen years old, and I was on Balboa Island, as the guest of a school pal. I sat with her on the sand, by the hour, watching the sailboats with a keen sense of enjoyment. I have always loved sailboats and still do, and I find them much more exciting than motorboats. But it is *watching* them that I love—not sailing in them! It was fun to see them dip and tip at times, and when one rolled over—which was frequently—no one got alarmed. The occupants usually were clad in bathing suits and could swim like fishes. And although it might take a while to right them again, it was all a part of the play. However, the time came when I was routed out of my safe seat and invited to take a sail. It would have been a breach of good manners to refuse my hostess. So, with trepidation, I embarked with their family and some friends on a small sailing vessel. All went fine while we stayed in the bay. But ere long we were out of the bay and on the sea. Then the rolling and pitching began. I was seasick, heartsick and fear-sick all at once! And that is pretty sick! As the waves got bigger I got sicker. I knew that each coming wave was going to be the one that knocked us over; and we would surely drown. Just one year prior to this time I had lost my closest girlfriend by drowning and the details of her shocking death were still fresh in my mind. I had lived them over again and again—the Sunday School picnic, the riptide and undertow, the sudden sweeping of her frail, lovely body out to sea, and the long wait before help could be summoned—help which came too late. I had seen her body, swollen and discolored, and wept with her parents. Then, clad in our white Junior High Graduation dresses, I, with others of her friends, had attended her funeral and seen her buried, just one week from the day of our graduation. As our boat tossed and turned that day, all this was vivid in my mind. I had felt sure that I was going to join my darling Rhoda in death—and by drowning too. Teenagers can be very dramatic! Up and down we went for a seemingly interminable hour. The problem was that having got out into the ocean, our navigator wasn't quite sure how to reenter the channel that led back to the bay! (Today an excellent breakwater makes it easily negotiable.) I surely made my peace with God that day and gave up all hope of getting back to land. (Yet, with shame, I now realize that our forefathers crossed the storm-ridden Atlantic and landed at Plymouth Rock in a vessel not very many times larger than the one we were in!) This early adventure, from which I returned unharmed, evidently left a deep mark upon my subconscious mind, and I must have relived it some deep way that night on the Nile.

I couldn't help but notice that our faluca seemed to be drifting very rapidly northward, and that the little dock from which we had embarked was nowhere in sight, as we begin to draw near the other bank. I also noticed that it took about three times as long to return as it had to cross over to the island. Or was it that I just thought it did? Had fear dimmed my sense of time? I recall sitting very tensely, holding on with both hands and evidently looking very sick. Dwight thought I was seasick, and I was content to let him think so. It helped to cover up my fear-sickness which was far more deadly.

I didn't realize that night why we were not getting to the right place on the river. But I have since learned that in Egypt the wind blows always from the north. Hence, when the wind is up the faluca tends to go south. However, when the wind dies down, the current, hastening toward the sea, carries it rapidly northward. The faluca however has very good tacking ability—hence it is preferred for use on the Nile above all other small boats. Now tacking is something I know little about. So, when our Nubian suddenly began his edge-walking "tightrope" exercise again, I was concerned to hear him say that since the wind had died down, he would have to do some tacking. I guess for a moment I thought the boat needed repairing. He had asked the heavy sister to move again—and sit solidly in the center. Again he worked away at the sail and got it shifted and we went through the tipping and rolling process. When we were righted again, we all sighted with relief. But this was of short duration, for the faluca headed rapidly out into mid-river again, leaving the shore behind so fast that I couldn't keep still.

"We surely aren't going to make another crossing, are we?" I cried.

The dear Nubian, whose name we knew then, but I cannot recall now, assured me that this was all a part of tacking, and that we would now have to take a zigzag course, in order to get back to the dock. We were already "zagging" enough to suit me. And the thought of zigging was just too much! I was about to burst into tears. But everyone else laughed and made a huge joke of it. Again I felt that I had lived through this experience before. But when and where? It is said that in each new sorrow, old sorrows tend to revive and we experience them briefly again. It is much like a wound that is reopened, or a scar that is always tender and sensitive to new injury. I believe this is true of fear too, and that the enemy has a way of reviving past fears when present dangers threaten. For many years I have waged strong warfare against fear in every form, knowing how deadly it is, and how I was once so enslaved by it in a thousand and one ways. Most of the time I am free from it. But I realize now that the enemy made a strong attempt to revive fear and bondage in me all during my stay in Egypt. Hence he revived past fears and linked them with my present ones.

Thus I must have relived, in a subconscious way, another terrifying hour I once spent on Emerald Bay at Catalina Island; I was about 27 years old. I had just been through a time of great spiritual crisis and blessing. In the midst of this I had rededicated my life to the Lord and was living in a high state of communion. With another couple, Dwight and I went to a four-day Bible conference. One afternoon while the men were all engaged in some recreation, my friend, Vivienne, invited me to get in a rowboat with her and ride around the smooth clear waters of the bay. She was a good swimmer and rower, so I got in the boat, disregarding my usual fear of such activities. For a while we enjoyed ourselves just floating around. Then I noticed that Vivienne was looking a little strained. Without our realizing it we had drifted quite a distance from the shore out toward the place where the bay merged with the sea. And the tide was going out! She asked me to take the other oar and pull with her. And although her voice was calm, I knew that she too was very frightened. After about 15 minutes of pulling with all our might, we both saw that it was hopeless. And I guess we must have called on the Lord in one accord. About then the lifeguard saw us and set out with another young man to rescue us. We both kept pulling as hard as we could, but the awful realization that he was going to have to transfer from his boat to ours filled me with terror, for already we were tossing roughly. I can still remember the tense and precarious moments when they finally reached us. They too were Christian boys, and we all were really looking to the Lord. I truly felt that day how easily we might have been swept out and on into death!

Such a remarkable deliverance should have increased my *faith* instead of my *fear*! Truly God has wrought great deliverances for each of us. Yet our human nature is more prone to dwell on the negative than the positive. I have felt led to relate these incidents because the Spirit desires us to be cleansed of past fears as well as present ones. As long as they lie deeply embedded in the mire of our subconscious mind, they can be suddenly resurrected at our weak moments. However, when the Spirit searches them out and reveals how faithfully God did watch over us and deliver us—even from dangers we encountered through our own foolishness—we can give God praise and transmute these experiences to vital testimonies of His power and goodness. Until I relived this night in Egypt and allowed the Spirit to search out my mind, illuminating these things that have happened thirty and forty years ago, *I had not fully realized how faithful and how great the Lord's power had been in those earlier days.* I had never given Him a full measure of praise for these rescues. However, if they come back to my mind again now, I am sure I shall be able to magnify His power and not the power of the threatening waters.

On that night in Egypt I was not able to do so. I only knew that I had a terror of being swept toward the sea, even though it was a considerable distance from Cairo. To add to my dismay, the little lantern on our faluca kept flickering and acting as though it were running out of oil and was about to go out. None of us had thought to bring a flashlight! Even that tiny light was such a comfort that the thought of its being extinguished was agonizing. And I kept wondering if it symbolized that my little light too was about to go out.

During all this time the "tacking" process continued. Now in case any of you are as ignorant about sailboats as I, you might be interested to learn that tacking is: "To change the course of a ship by shifting the tacks (ropes) and positions of the sails and rudder." And this reminds us of an old saying, "It's the set of the sail and not the gale that determines the course we take." Again and again we headed for the shore. Each time our Nubian "cat-walked" along the edge of the faluca with grace and balance, and the sister shifted her position. And each time we zigzagged a little and got nearer to the dock. I felt the surge, the splash and sometimes a little spray from the Nile—and began to have faith enough to believe that my baptism into the Nile was going to be of the Methodist type after all—only a little sprinkling and not an immersion. And I was content to have it so!

By this time we had learned a little of the troubles of our boatman. He had several children and two wives. He was Moslem. He worked long hours to support his children. He had abandoned his first wife when she refused to keep house—but he still supported her. His work was hard. His hours were long and his pay small. I can't recall that one of us said a word to him about salvation or Jesus—except that he knew that we were on a Christian tour. Fear had eclipsed all thought of others from my mind, and had enslaved me during the entire boat ride. My one thought had been to escape, to flee from those dark waters.

As I look back upon this experience I can see many lessons in it. I know that all Israel feared Egypt and sought to escape from its power. There is a strange force of darkness there which is difficult to understand, but it must be similar to that which is encountered by missionaries throughout Africa and Asia. Yet the Lord had given me the word: "My heart shall know no anxious fear." And fear is truly a form of unbelief. I thought of how Moses had been committed to these waters and had been drawn from them by the hand of God, through an Egyptian Princess' hand. What faith his mother must have had! And yet it was a faith born out of fear . . . fear for his life. I remembered how this very river had been turned to blood, by the hand of God. Yet I had feared to sail on it, not fully trusting this mighty hand! Again He comforted me, for did not the disciples cry in fear on the Lake of Galilee, and they were experienced fishermen and boatmen. But surely Jesus must walk on the Nile too! And He is the Master of the waters of Egypt, as well as of Israel. I realized again that I was not very willing to meet death by drowning; yet if we are truly dedicated, we should not have such fears. So, I sought to be cleansed from them—and from those past fears that still arise from earlier experiences to haunt me now.

My saddest recollection about this night is of the moment when we finally reached the dock and the Nubian took my hand to help me get off the faluca. For an hour or more my life had been completely in his strong black hands, and he had delivered me safely to shore. Likewise, for that hour, his precious soul had been in my hands, and I had done nothing to secure its eternal life! I have since tried by prayer and faith to secure his salvation. But I shall not likely ever see him again on earth. I have been a very poor missionary indeed!

This "candid-camera" record of my baptism in the river of Egypt is surely lacking in romantic beauty and inspiration—indeed it is utterly inglorious. Yet I am sure that the Father did permit, if not directly will, this extremely trying and revealing experience. It proved to be a fitting prelude for all that was to follow—for in Egypt it seemed that we were "in jeopardy every hour." And I was tempted to be fearful a dozen times a day.

That very night I had to make a definite stand in the Lord and refuse to spend my time in Egypt fighting fears. Fear and faith just do not walk together! And "that which is not of faith is of sin." I realized by now that I was going to have to move very carefully and prayerfully indeed, and that I must at all times be filled with faith, thanksgiving, and praise, if I was to glorify the Lord during my remaining two days in Egypt.

My baptism in the Nile had another very surprising effect upon me. Instead of disliking this great river, I found that I had a wonderful sense of being identified with it and loving it. Of all the rivers I saw on the trip it thrilled and satisfied me the most. And even now when I think of it there is a warm glow of pleasure in it. I someway feel that I too, like Moses, was drawn from its waters. And some of you have told me of your particular love for it. Certainly if I was thus baptized in it, each of you were too. So with me you can sing an ancient song of David, who never saw the Nile, yet prophesied: "Then the channels of waters were seen . . . He sent from above, He took me, He drew me out of the great waters."

IN COLORFUL CAIRO

idnight . . . a balcony . . . and a peacefully sleeping city. What delight this happy combination had brought to my heart in Jerusalem! But midnight on a balcony in Cairo was most disturbing. I wanted to close the French door tightly, and shut out all the sights and sounds of the city, if that were possible. For one thing, of course, Cairo was not peacefully sleeping, even at midnight. In fact, it was not sleeping at all, or even yawning. The din of horns, the voices of men and women, the blare of music on radios—and who knows what else—all blended in a wild sort of mélange. It was not nearly so loud and raucous as the cacophony of the traffic rush hours—but it was distinctly audible, even from our 15th story lookout station. However, as I have said, it disturbed me to look out upon it, for Cairo is a disturbing, restless and somewhat formidable city. But intriguing! Most intriguing!

It is said that writers have exhausted themselves in several languages trying adequately to describe Cairo so that their readers may capture its atmosphere and be able to picture it for themselves. Most of them have admitted that it just can't be done. And I am ready to agree with them. The nearest one can come to an adequate description is that, "It is a combination of Damascus, Baghdad and Calcutta, with a touch of Paris thrown in." However, since you and I are not very familiar with any of these cities, it is not likely that we can imagine accurately what Cairo is really like. On this, my first night, I felt that I was in a strange, unreal sort of almost nightmarish dreamland, among a strange people. And like Joseph, "I heard a strange language that I understood not." By this time Arabic was beginning to get on my nerves a little. It is said that it is impossible to whisper in Arabic, and that the most peaceful conversation always sounds like a rousing fight. Egyptian Arabic is harsher, more guttural than that of Jordan, and much more predominantly spoken. Indeed we seemed to hear nothing but Arabic on its streets. It was odd to find ourselves in the position of being foreigners, unable to understand what was being done and said around us. It was also strange not to see any other white-skinned people, save those of our own little band.

Now, about now will you may begin to wonder how I had acquired such a close contact with Cairo in such short order. You may have imagined that after our devastating boat ride we hastened to the hotel and retired to recover. But such was not the case at all, though I had innocently supposed that it would be thus. The evening had grown quite late, and our weariness was by now sheer exhaustion, when we finally were settled in a taxi. I was relieved to leave the Nile behind and head for our hotel. I resolved to shut my eyes and refuse to see all the near collisions and calamities we might encounter in the short ride through the city. I would just keep thanking the Lord and praising. Then, suddenly, we would be at the hotel and I could go to bed. Sure enough, the taxi made a hair-raising but safe run through the streets and then wheeled to an abrupt stop. Our smiling guide rushed to the door to help us out. He was obviously delighted to see us again. The only thing wrong was that we were not at the hotel at all! To my amazement I saw a large, brightly lighted shop, inside of which members of our tour were milling around, obviously shopping. My patience began to crack under the strain. Oh no! We surely weren't going "souvineering" at this hour of the night! I began to protest. "But, Madamé," our guide said, in his persuasive, unctuous manner, "we have made special arrangements to open the store as a favor, to accommodate you, so that you may have opportunity to see the very finest of gifts and souvenirs in a leisurely way." I tried to tell him that I was not at all interested in going shopping at such an hour, particularly after the kind of day and evening we had gone through.

It was useless to protest. We were obviously trapped, and had to make the best of it. I found a sense of indignation rising in me. What right had this Arab to just take us over and force us into his store. (No doubt he had a financial interest in it.) In Jordan we had seen a bit of this type of thing, but it had been done in such a frank childlike manner that we could laugh it off. But in Egypt it was different—it was extremely professional and outright nervy. We were placed in such a position that we were almost forced to buy—in order to get back to our hotel. I got stubborn about then, and said that I had no intention of buying *one* thing that night. I was just too tired. I looked in vain in the shop for a chair on which to sit—but of course, none was available.

I very nearly just sat down on the dirty floor—I was that "far gone." I even considered trying to faint or collapse, so they would have to carry me out or something. But, of course, I did nothing of the kind. Dwight, meanwhile, was explaining that he could not shop, for we had no Egyptian money. But this was no excuse at all. They were willing, delighted even, to take the good old American etchings of our Presidents—or travelers checks. I noticed that he too was beginning to get rebellious inside. But others of our companions were shopping away with apparent enjoyment.

I found it hard to accept or understand the great interest Pentecostal Christians had in shopping, no matter where we went. Apparently many of them had an almost endless supply of money with them, and they loved souvenir hunting. One very devoted sister collected an expensive bracelet from each place we visited, and usually wore them all at once! I was tempted to feel critical about this, but knew that I shouldn't. So, I just milled around inside and outside, looking at the people on the streets, and sensing more and more how strange everything was. Or could it be that it was we who were strange? The ordeal finally ended and we had another dashing ride to our hotel. Just before reaching it, I observed a huge statue, nicely lighted up. This proved to be Rameses II. And it is now believed that he was the Pharaoh who withstood Moses. As I looked upon his huge, arrogant face that night, I felt that someway he characterized modern Egypt also, and Nasser, with his hatred of the Jews and the ambition to dominate the Arab world. I was somewhat shaken in my soul, sensing that I was indeed in Pharaoh's Land—and that the enemy was both strong and very active.

When we stepped out of the elevator, we noticed that the lobby was almost empty. All the luggage of our protesting companions had been removed. And we wondered what had happened to them all. There was a new deskman on duty, and he seemed not to understand English at all. We had a hard time getting the right key. Standing off to one side was a man who was by now rather familiar. He had been there in the lobby each time we had passed through. There was something rather attractive about him. But his quizzical, faintly amused attitude was puzzling to both Dwight and me. He spoke English fairly well, and his nationality was difficult to ascertain. He had a way of drawing us out and getting us to talk, while being very discreet himself. I engaged him now in conversation, while Dwight argued with the clerk. And I wondered why he tried to find out where we were going next on our tour. I told him frankly that we were going to Jerusalem. Then, noting his raised eyebrows, I had quickly added that it was in Jordan. Actually, up to that moment I had not fully realized that, of course, we were going back to Old Jerusalem. One could not possibly fly from Egypt into Israel! I can't tell you how happy this made me. Feeling so heartbroken at leaving there, it now thrilled me to realize that, if all went well, we would be back in Jerusalem in three days! However, I continued to feel odd about this man. Dwight was sure that he was a detective or some special agent put there to keep tab on us. On us? How very funny! We are so harmless! Then I realized how often we are suspicious of foreigners, just because they are of another race. I, for instance, as a child, used to fear Chinese very much, and also Negroes. Yet Americans are the worst criminals of all! Hence we went to our room that night, feeling a sense of high adventure, intrigue and international conspiracy. It was all very funny and also a little dismaying. No wonder I wanted to shut out all the sounds and thoughts of Cairo!

In spite of the funny beds, we slept soundly—the sleep of sheer exhaustion! Dwight, as usual, arose with the dawn. And it comes much too early in Egypt too! He was all for bathing while the water was hot, and then exploring the neighborhood. I was all for going back to sleep—feeling utterly unable to face the day. But the chorus of Cairo was now swelling in volume, and every little toot reminded me that it was time to "rise and shine." This, our first full day in Cairo, was going to be an exciting one, so it was fitting that it began with a little private flood. Dwight ventured into the shower and finally got the hot water to run. It seemed to run in no definite direction, but splashed high and wide all around, and turning it off proved difficult. By the time he succeeded, the water was standing on the floor about four inches deep. Of course, since the floor was cement, there was no damage, unless it had risen an inch or so higher and overflowed the bedroom. However, it did seem like a little moat, through which I had to wade to get out the door and to the hall lavatory. We concluded that the drain was stopped up, and that perhaps in time it would run out. The thought of summoning an Arab bellboy was more than we wanted to face. So I realized that I had best skip the shower and rely on the good old towel bath, which I did. And I managed a little laundry before it was time to go to breakfast.

Dwight, meanwhile, returned from his little sortie into the streets, feeling quite frazzled. He was simply amazed at how many people tried to sell him something or beg from him. He said he had been jostled about most rudely. And he also warned me not to go out on the streets alone at all. "We are just curiosities," he said. "And they think we are rich." And I could see that he was annoyed. We were among the first to go in to breakfast. And we sat down at a table with some colored sisters. One of these was especially pleasant and spiritually bright, I had noted. And I felt the need of a little fortifying. We talked of the Lord, and tried to keep everything cheerful and to His glory. But I noted that some of the others were already talking about souvenir bargains and what they intended to buy. Later more of our companions came in. But the "dissenting delegation" was still missing. So we supposed they had left the hotel. Breakfast was frugal. The coffee was odd tasting. The eggs had to be ordered and paid for extra. Since I don't care for eggs anyway, I stuck to the hard roll and coffee, and ate a little jam. I was glad that I could go to my room and eat protein pills and vitamins to make up for the lack. Thus fortified, I straightened up our room and prepared to lock my bag. We had only about five minutes now to get downstairs to the bus. But try as I would, I could not lock the suitcase, nor could Dwight. Inside it were some of my costume jewelry and personal things, as well as all the mother-of-pearl pins and jewelry I had bought in Jerusalem. I knew that these things could easily be stolen and sold. So it was a trial not to be able to lock the bag. There was no time now to take everything out and put it in Dwight's bag. So I just had to walk off and leave it, praying that the Lord would protect it.

When we finally got the elevator and descended, we were late. But the bus was nowhere in sight. So we joined the group standing on the hot pavement and waiting. Already the Egyptian sun was blasting away! The people swarmed all around us there in the crowded street. And the peddlers were indeed numerous and persistent. Children boldly approached us, crying, "*Baksheesh!*" And the rest eyed us with interest and curiosity. We tried to be patient and friendly, but were relieved when the bus finally arrived. It had already circled around, picking up others of our tour from two hotels. The peddlers now approached us at the bus, reaching inside and putting their wares in our laps. One colored sister entered into a lively session of oriental bargaining. And several bought from them. Those who didn't soon got tired of it—so we closed most of the windows and were uncomfortably hot.

Meanwhile, we waited. But we knew not what for. Finally one of the men told us that we were waiting for the "dissenters," who had gone to the U.S. Embassy. We could not leave without them. The passengers then took sides whether or not this group should have complained about the hotel accommodations. Most of them felt that they should have just made the best of it. But these, we found out, were in hotels much better than ours. We had apparently gotten the worst one available! Talk spread about this one and that one being ill. Dysentery was discussed, along with several other diseases, said to be prevalent among tourists. And it was alarming to hear that already several of our number were quite ill. Fear and confusion is very contagious, and it spread insidiously among us. I tried hard not to get under it, but to keep praising and thanking the Lord. And I was glad when I heard one brother say, "Well, what's the matter with all of us? After all we are Christians. We do have faith and we have the Lord. Why do we have to be so afraid of disease?" About that time "the delegation" arrived. None of them apologized for keeping us waiting. They merely got on and began talking all at once. The leading sister finally took over. She said that they had not been able to do one thing about changing hotels. The Tour Representative had never even been found, and since they had no passports they couldn't go anywhere else anyway. They had someway suffered through the night. Early that morning, several of them had gone to the U.S. Embassy. But this too had proved fruitless. They had, however, secured some dysentery drugs for the two or three who were ill. They had also been told that the food at the Cairo Hilton was no safer than anywhere else. A nurse had warned them to eat as little as possible. She also said to avoid meat, which was very dirty, and bread, since it was freely handled by many. Already milk and milk products, raw fruits and vegetables had been banned. Now, in addition, we were to avoid meat and bread. Just what did this leave for us to eat? And of course we were to avoid the water. The protests began to rise like a chorus. It is obvious that almost everyone was upset. I don't know how this would have been settled, had not Leta Mae Stewart suddenly appeared and boarded the bus. We hadn't seen her since we parted in N.Y. And just the sight of her was reassuring. Smiling, as always, and very calm and charming in manner, she addressed us all with quiet authority. Someway she managed to calm our fears and to assure us that usually only a very few tourists became ill in Cairo, and this could happen anywhere. She urged us to be temperate and to have faith. Her confidence spread like a healing oil over my heart, and I resolved not to be moved by fear, regardless of how many got sick.

Eventually, of course, we got started on our tour. We were a long ways off schedule, and quite worn out already. It was not an auspicious way to start the day. In spite of all these disadvantages, I found the short tour through the streets of Cairo both exciting and enjoyable. Cairo by day was not at all formidable, but most interesting. We learned that it is not an ancient city at all, as I had supposed, but was founded in about A.D. 968. Before that time Alexandria had been the great city and capital of Egypt. Its original name was *El-Kahira* (Ka-HAIR-a), meaning "the Victorious." Later it was corrupted into Cairo. One writer has said: "Nothing that the traveler has ever seen is quite like colorful Cairo. And he will never be able to forget its panorama of human life which never ends; not even at midnight; the hundreds of thousands of lives which nothing seems ever to perturb; the glow of the city in the sun; the shimmer of the Nile, and the pyramids in the background. One wonders if there is another scene in the world so vast, so solemn, so thrilling, as this great city on the edge of the desert, in the glare of Egyptian noonday. Was there ever such a blending of color, or such a mixture of peoples? Hawks fly past you as you walk in the street, water buffaloes draw carts and plows, white donkeys and black ones bear their burdens, women hide their bodies and their faces in black veils, and children play everywhere along the streets."

Yes, Cairo is indeed colorful and exuberantly alive—that is, until one reaches the stately Museum. Then, passing through its massive doors, one is ushered solemnly into a great and majestic Temple of the Dead. I was hardly prepared for this transition. And its effect on me was both impressive and depressing. Outside everything had been very gay. By this time we had all cheered up and were enjoying ourselves. The head guide of those who were delegated to conduct us in Egypt was there to welcome us. He was dressed in the long flowing robe and turbaned in headgear of the native Egyptian man. And since he was a huge and jovial man, he was a rather splendid figure. "Just call me Daddy-O," he told us, delighted to display his knowledge of American slang. He posed for his picture again and again, and Dwight snapped one of me with him. He conducted me in person to the beautiful lotus pool. Not too many were blooming, but they were lovely. He said Americans always call them water lilies, but they are truly the lotus. (However, the real Egyptian lotus, for which they are famous, is not of the water lily type. They grow up very tall above the water. It is this type of lotus that Sister MacPherson brought from her Holy Land tour, and these now flourish in Echo Park. Each year they bloom there in August, and they are a splendid sight indeed. We learned that they flourish in the famous gardens and pools along the Nile in the Delta.)

Another guide led us on and up the steps through the great doors. The first impression was that of delightful coolness, after the burning heat. But soon I felt almost chilled, as though I had entered a tomb. Egyptian statuary is massive and overwhelming. It does not have the grace and beauty of that of Greece and Rome. Almost everything in this Museum was ancient, very ancient. Anything less than two thousand years old is "new" in Egypt! Since all the interest and concern of the ancient Egyptians centered about death, their art also dwells upon this theme. Their many gods, most of them with animal heads, are revolting to us. As I looked upon them I couldn't help but be thankful for Islam. The Moslem, at least, seeks and worships a god who is a creator, not a creature, and who can never be depicted in any form, let alone in that of an animal. Hence such gods were banished from Egypt.

Some of our companions became bored within a few minutes. They grew tired of walking, and began to sit down to rest or to wander around. They paid almost no attention to the guide. He tried so hard to hold our interest, that I felt sorry for him. And I think he found us quite uncultured. Egyptology, is, after all, a great and thrilling study to most educated people. Because of the lack of rain, the art treasures of very ancient days have been preserved in Egypt's dry sands. And here one comes into vital contact with the civilization of thousands of years ago. It is almost overwhelming to consider the antiquity of such treasures. Our civilization in America has existed only a few hundred years!

I did my best to absorb and learn all I could that day. But I could not retain it. I hope eventually to be able to study this subject, at least in a cursory way, for I am convinced that it is of vital importance to every Bible student. After all, Egypt and her civilization has a very prominent place in the Bible.

I am sure that the most thrilling display I saw was the Tutankhamen Tomb treasures. A whole section of the Museum is devoted to these. The discovery of them took place in our own life time, in 1922, and we can perhaps recall reading about the great importance attributed to it. We may recall too that everyone who entered the tomb died within a fairly short time afterward—and that the Egyptians attributed this to an ancient curse. Most of the tombs discovered recently had been plundered in the long ago days. But most of Tut's treasure was found practically undamaged. His tomb contained hundreds of priceless objects—chests, caskets, vases, chairs, stools, thrones, statues, chariots, figurines, jewelry and precious stones. These provide an unparalleled revelation of the supreme artistry of the Egyptians living in that period of time (1358–1350 B.C.)

About the time we began to view these treasures, a distinguished party of Yugoslavians entered also—perhaps some governmental representatives. Strong lights were thrown on portions of it, so that pictures could be taken. This enabled Dwight to get some good shots too. And how the gold and precious stones glistened in this floodlight! The beautiful shade of Nile blue, which I dearly love, and lapis lazuli, were used in almost all this artwork. And I could not help but marvel at the massive gold and blue mask which covered the head of the King's mummy. The face of this is an admirable portrait of the King, and it is indeed handsome. The Queen was beautifully portrayed on the back panel of one of his thrones. The actual tomb of the King is at Luxor. But his mummy was placed in several sarcophagi, one within another. And each is of solid gold, encrusted with fabulous jewels. Alabaster in most beautiful shades was also used for some of the articles. It seems likely that no one in the world ever had so extravagant or rich a tomb as he. Yet, we do not know, for many Egyptian treasures still lie buried in its sands. I could not help but contrast his fabulous tomb with the simple rock tomb of our Lord. I can readily see why the Israelites were scandalized by all this adornment of tombs. Strange Egypt, letting its living starve, and clothing its dead with riches!

(Most of the Tutankhamen display is now in the United States! For the first time it was allowed to leave Egypt on a tour, for the purpose of raising money to help preserve some of the temples which would be inundated by the new dam. If this comes your way, by all means try to see it.)

I am sure I left the Museum with a new sense of the beauty, artistry, grandeur and glory of the great civilization of Egypt, the world's greatest country for a long period of time. It seemed so odd then to go out into its streets and its present state, realizing how very backward it is among modern nations. The Protestant Christian nations are the most progressive on earth. The Islamic are among the most backward. This alone should speak clearly to all who can read history aright!

It was time now to return to the hotel and to "enjoy" lunch, we were told. Then, oh joy, we shall be off to the pyramids! The very thought of it was overwhelming. We were a tired and hungry lot, and would have given a lot to have a nice "safe" American lunch about now. But even experienced travelers say that they can't think of one place in Egypt to recommend for dining. This seems strange, when one considers that the French have had such an influence on Egypt. It is said that the British ruled, but the French taught—mostly through Catholic missionaries. Therefore it is the French who have had the greater influence. About the only indications I observed of this were the French signs in the Museum and most public buildings. Arabic, English and French are used. And among the city people it is rather widely spoken. How soft and musical it sounded, as contrasted with Arabic! It seems to add a special charm to Colorful Cairo.

A CAMEL-RIDE TO
THE PYRAMIDS

T he seven wonders of the ancient world have all decayed and disappeared, with the exception of the great composite Wonder of Egypt—the pyramids and the Sphinx. Majestic, impervious to time, monumental on a colossal scale, these stand today in the midst of the land of Egypt, just as they stood in the days of the patriarchs. It is possible that Abraham saw the Sphinx when he visited Egypt, for it is the oldest human structure on earth, having been in existence for at least 5,000 years.

The Great Pyramid of Cheops is the most impressive of the massive monuments constructed centuries ago. (Modern builders are still baffled trying to understand *how* it was so perfectly engineered and erected!) Originally this was 483 feet high. However its height today is only 451 feet, since the peak was removed and the stones used for building much of the old city of Cairo. It still contains 2,300,000 huge stone blocks, many of which weigh two and a half tons each. Dr. Field said that these would suffice to build a wall one foot thick and six feet high from New York to San Francisco. Imagine this! I find it hard to grasp! Some of the stones weigh as much as sixteen tons. Yet all are held together by a layer of cement no thicker than a *hair*! Herodotus said that it took *one hundred thousand men thirty years* to build it! It is no wonder that this great pyramid and the lesser ones, along with the incredible Sphinx, are the "number-one" tourist attraction in the world. Myriads of scholars, students, Bible lovers, curiosity seekers and just plain tourists have made a pilgrimage to Egypt to behold the only remaining wonder of the original seven.

The Egyptians are very proud of the many important antiquities that have been preserved in their present-day "insignificant" land. And, to be sure, they tend to take full advantage of the unsuspecting traveler who wants to view them. Since returning home I have learned that it is an Arab tradition that any traveler who comes to their land is sent by Allah. He is indubitably rich, for otherwise he could not travel so far. He is also curious, and perhaps unwise, preferring to spend his money on travel, rather than to provide more abundantly for his family. Therefore Allah expects him to be generous to all Arabs, and to reward them for their hospitality, which is world-famous. If the traveler is generous, then Allah will protect and bless him. But if he is stingy, he will surely bring down the wrath of Allah on his head, and will suffer because of his lack of liberality. Considering this tradition, I can readily see now why all the Egyptians were so anxious to have us be generous: it was for *our* sakes, so Allah could bless us! And here we all thought it was for their own benefit!

In view of the impressive age and dimensions of the pyramids, it would seem that one should approach them reverently, pondering the mystery of their origin, and recalling the great personages—kings, queens, saints, conquerors, and perhaps even Christ as a Child—who have visited them throughout the centuries. It has been said that it is good to approach them first at night, when the moon and the stars shed upon them an unearthly glow. How I would have loved to have done so! Next best would be at sunset, when twilight casts its ages-old spell around them, adding an aura of mystery. At dawn would have been another evocative time, it seems to me, though I guess the sun comes up very suddenly in Egypt. The worst possible time to make the pilgrimage is during the middle of the day—or at least so it seemed to me. There is something blatant and merciless about the noonday in the desert land! Yet it fell to our lot that shortly after lunch we were loaded on the buses and told that soon we would see the glorious pyramids.

Try as I did, I just could not seem to get myself into the right mood to visit them and the "Sphinkus," as our guides fondly referred to it. Sitting jammed in an uncomfortable seat in that hot bus, I tried to tune out all the irritating noises and sights around me, and to praise and thank the Lord and seek His guidance. But He seemed far removed from His restless, complaining little flock of Pentecostalites. No peaceful noonday rest for us! I have already described in a previous portion the exercises and difficulties of our morning at the museum and our noonday repast at the hotel. A number of small additional happenings had served to upset us somewhat, none of which is worth reporting. But it was *not* a small thing to me that there was still no mail from any of you! I felt so in need of a touch, a transfusion, about then! I was a stranger in a strange land, and I sorely needed to get my bearings. How I longed for some sort of outward communion! But I had to content myself with taking this communion by faith, apart from all awareness. I was sure in my heart that you had written, so I knew it must be the enemy who was keeping your mail from me.

It was also disconcerting that we would not get our passports back. Without them we could not exchange any money. Already our American dollar bills were dwindling—in fact mine were all used up, since that morning I had purchased some souvenirs from street vendors. We had larger bills and travelers checks, but could not turn them into Egyptian money. (Our silver they will have nothing to do with, for some reason.) So I had to keep telling the vendors who were pestering us at the bus, sticking their merchandise through the windows, that I had no money. And whether I said it in English or Arabic, they refused to believe it and shook their heads, acting like this was incredible! They had an annoying way of dropping their souvenirs into your lap and refusing to take them back, meanwhile screaming at you, "Please, Madamé, one dollar, one American dollar, please Madamé, etc." At length several of us grew so frustrated that we rudely shut the windows in their faces. But others kept haggling with them, and the atmosphere was hot, tense and utterly distracting.

It seemed like an endless wait before the bus finally took off. The guide was trying to round up all our missing members—some of whom were too ill to go with us. I can't imagine why the bus was still so crowded without them, or where we would have put them had they appeared—but I felt pained for them, nevertheless. I had already learned, of course, the rules about bus-seat priority for our Tour: those with cameras had first rights to the seats by the windows; the larger the cameras were, the higher their priority. And if one had a movie camera, it was higher yet. (One dear brother managed to cart along two cameras everywhere, so he was "top man.") The only way I got to sit by the window was when Dwight permitted me to do so, as he usually did. Since he was a "large-camera" man, he had window privileges. I had also learned that some of our number had no intention of sitting toward the rear of the bus where the fumes were quite noticeable, and where some of the seats were crowded and uncomfortable because of the wheels. So we usually ended up back there. Another most annoying thing about such buses is that the windows are very wide. Since everyone was hot, those in the seat in front of us opened their windows. However, the rush of wind bypassed them, hitting us full in the face and on the throat—a sensitive spot for me. Those behind us, on the other hand, insisted that we open our window too. So this put us directly in a strong draft. I was already beginning to feel a bit hoarse and tight in my throat. And I dreaded taking a cold! My hair, of course, was just wild most of the time. It stood almost straight up. I learned to just endure it, and cover it when we finally stopped, putting my hat on at once. I could not wear it in the bus, because it just might get in the way of someone's picture shot. It was indeed trying to be told every few minutes to duck down, move over, or get out of the seat entirely, while the bus stopped for special shots. Now all this, of course, should not trouble a real saint, and maybe you would have been utterly unmoved by it. But I was not feeling very saintly about then, and I was keenly aware of these and other annoyances. Hence, perhaps you will understand that it took a real effort to try to praise and thank the Lord and be in any suitable frame of mind for the great event which I hoped lay just ahead—a memorable visit to the pyramids.

The bus ride was short, for the pyramids are located only a little distance from Cairo, in a suburb of the city called Mena. The highway leading to it is rather wide and attractive, following the Nile, and lined with modern homes and buildings. There were some attractive, if not lavish, gardens. I was most eager to know the names of the flowers, especially of one orchid-colored blossom which was in full bloom on many a bush. But the guide seemed to know nothing about flowers. (And this I found to be the case in every land we visited.) I recognized the roses, lilies and bougainvillea—which was in evidence everywhere. I noted that there were only a few lawns of any size. If one saw no more of Egypt than this road, it would be easy to think that it is fairly modern and prosperous. But not far from the highway, we began to notice the tiny mud huts, and as we neared the pyramids, there were children begging along the road, hoping we would stop for a picture and give them some Egyptian coins. They looked half-starved, and my heart was torn with pity for them. But alas, I had no coins. (They throw away our silver, thinking it has no value. Perhaps they cannot exchange it.)

Suddenly, we heard a chorus of shouts and strange cries, and we saw a lot of camels, each laden with a tourist, being led by natives in Arab dress. And then— there they were—the Sphinkus and the pyramids, just a little distance down the road! I got only a peek at them, for all the cameras suddenly went into action. The bus stopped, and the men rushed off first. The sister photographers were next.

Meanwhile, I held back, ashamed of our behavior and determined to be ladylike and polite. On one side I could see a very old English-type hotel, swarming with guests. And all around us there was much noise and confusion. It was apparent that we were going to be well-surrounded during this part of our tour.

Our guide had already announced that this was where we would get on our camels for our ride to the pyramids. Those who did not want to ride a camel could remain on the bus, and it would be driven to the pyramids, where all would be reunited. I knew that this camel trip was provided in our tour, and I had felt a witness that I was to take it. Very vividly indeed I recalled how the Spirit had dealt with us at various times about Rebecca and her long, rough camel ride to meet Isaac. At one time the book *The Camels Were Coming* had been a great blessing to me. Thoughts of it flashed now through my mind.

I also recalled the time Arthur had a strong "Rebecca" dealing about me, shortly before the Lord moved us by car—not by camel, thanks be—to the Mount. Arthur had tried to locate a camel for a demonstration. Failing this, he got a horse, and came riding over to the house we were building in Inglewood, clad like an Arab. I'll never forget how surprised Dwight was when Arthur rode up on the vacant lot next to us! I knew he was coming, but I was even more startled! To be true to the scriptural pattern, Arthur insisted that I offer water to the horsey-camel, which I attempted to do, after finding and filling a bucket. It was sort of a messy affair. The horse was evidently playing camel, and they don't drink very often. Next I was obliged to get *on* the horse, a frightening thing to me. About that time Arthur felt that the sign had been performed, so off I got and in he came, bearing gifts and Scriptures for me. It was very precious, after all, and the Spirit sweetly moved on several of us who were present. From that time on, I had always felt that someday I would take a literal ride on a camel. Now it gave me a real spurt of joy to recall that long ago demonstration, and to realize that I was about to take that ride—and in the land of Egypt by the ancient pyramids which even Abraham may have looked upon.

I was still in this happy mood as I walked across the road toward the camels, just in time to see Dwight riding off in the other direction! I called to him, and he someway heard me above the ruckus of protesting camels and shouting camel drivers. He turned and gaily waved, calling out, "See you later!" This, for some

269

reason, unnerved me. Some of the other husbands were helping their wives mount, obviously intending to make the journey together. I was tempted to feel chagrined that my husband apparently had no thought of me, but had abandoned me to the mercies of the Egyptians. And I was a wee bit scared, I guess. Then I remembered that the Lord had shown me ahead of time that though Dwight would go with me on the trip, I was not to look to him as a husband, but only to the Lord. If I was writing as a Rebecca, my Isaac would not be visible. But He would be at hand, praise Him! It was going to be hard though, I felt, to see Eliezer in any of those Arabs! And, after all, Rebecca did have some other familiar attendants.

As I watched the women mount, I was glad I had remembered to wear a full skirt. I was at the end of the line, so tried to wait patiently. About then, two donkey carts approached me, and their drivers began a good-natured competition to get me to ride with them. Each tried to outdo the other in luring me into his cart. From their manner one would have thought that they were offering to take me absolutely free, and that they would be heartbroken if I refused such kindness. I tried to make a joke of it, and resorted to Arabic. This only aroused their interests more, and soon I was really embarrassed at the attention being paid to me. I kept telling them that I was going to ride a camel, not a donkey cart. But they grew more insistent. About that time, I felt it was enough and abruptly turned and walked off, thinking it must be my turn to get on a camel. Then, to my dismay, I saw that every one of our tour had mounted and ridden off—and there stood only the guide, looking sheepish.

"I am so very sorry, Madamé," he said, "but there are no more camels . . . all are in use. I am so very sorry." He bowed and smiled most engagingly.

I was simply stunned! After all my anticipation, preparation and appropriate meditation, here I was stranded without a camel! The donkey-cart men took in the situation and renewed their "attack." And this was just too much! I fought to hold back the tears, and bit my lips to keep from saying unkind words. I knew that the ride was included in our tour, and that I had every right to a camel. But who can produce a camel out of thin air? Our guide was no Aladdin! So I tried to be gracious, though I could not quite hold back the tears. I slowly walked back to the bus. The women who were waiting were impatient to be on the way. And they thought any woman who wanted to ride a camel was a bit nutty. So I said nothing, but sat down and tried to call upon the Lord about it. In the midst of my rather feeble prayer, I was violently attacked with a temptation to be really upset with Dwight. Why had he been so selfish as to rush ahead, without a thought of me? I felt indignation mounting, and was weakly resisting it, when the guide suddenly stuck his head in the door of the bus.

"Madamé, Madamé, come quickly, please! For you I have a camel, thanks be to Allah." Involuntarily I shouted, "Praise the Lord!" And I fairly bounded off the bus. Now I was *sure* that the Lord wanted me to have that camel ride. He was my own Aladdin, and had produced one. I was in a rosy haze of gratitude and goodwill as I approached the mean, rather mangy looking beast who glared at me for a moment, then gnashed his teeth rudely and stared off into space. Camels are rather unfriendly beasties! But the driver's delight more than mitigated the camel's ire. He fawned and fussed and treated me like a queen. The camel was dressed in oriental fashion, and so too was the driver. It was colorful attire, but rather dirty and smelly, nevertheless, and I gently refused the offer of an Arab shawl, which he almost insisted I wear. I was already hot enough! The camel knelt for me—not in humility—and I got on without much trouble. However, the getting *on* is nothing in comparison to the getting *up*. One must hold on tightly and cooperate fully to remain in place. But once he was up, the camel's back was not too uncomfortable. I had read that if one just relaxes and flops in every direction, as the camel wobbles, one will then be able to ride with fair ease. And I found that this is true.

It was really fun to be perched up high on the camel's back, for I have been more or less fascinated by them, since I first saw one in a zoo, when I was a little girl. As the driver led him up a little hill, I could see an impressive pyramid just ahead, and I felt almost ecstatic with expectation. I could not recall consciously very much I had studied about the pyramids or the Sphinx, but I had deep impressions of them stored away in my subconscious. It was going to be a privilege to worship and commune with the Lord here, and make each of these an altar for Him. I felt almost rapt for a few moments, and I guess I showed it, for the driver suddenly called to me loudly, "Madame!" I was instantly grounded—shocked by the tone of command in his voice. How quickly he had changed from a fawning servant to a demanding Bedouin!

He appeared to be angry with me and to be accusing me of riding without paying; or so I thought. So I undertook to explain that we had already paid for our rides in our tour fee, and that our guides would settle with him. But I could not seem to make him understand. He kept demanding money, and I was embarrassed and tempted to be annoyed. One does not readily sacrifice a high moment to a mean little man—and this was just what he was acting like. Finally I asked him if what he wanted was a tip—and his face lighted up. He thrust out his hand and demanded it at once, meanwhile reciting a singsong story about how poor he was, how large his family, etc. I recalled then that we had been warned that these drivers would demand money and plenty of it. If one gave in readily and tipped them, they would then demand more. If one did not give in, they would keep insisting. In either case they could and would make your entire ride miserable with their demands. I had already decided that I would gladly tip the driver. After all, we are rich indeed, in comparison with these poor Egyptians.

In America we tip for all sorts of pleasures. So why not in Egypt? I didn't feel that it was Christian or kind to refuse the driver a tip, and I didn't doubt in the least that the owners of the camels took the lion's share. However, I was in a dilemma. I simply had no Egyptian money with me, and no American dollars. I did have larger bills and traveler's checks, but these I could not use. So I tried to tell the little man kindly all about it and assure him that my husband would provide the tip at the end of the journey. This made him so angry that he stopped, stamped his feet and gnashed his teeth at me, much like the camel. At which the camel too snorted and made an outcry and shook. I was a little frightened, and suspected that they were in league. (Maybe if he doesn't get paid much, the camel doesn't get much to eat!) I wondered what I would do it the camel suddenly bolted!

We were in a rather lonely place. Already we had passed the pyramid and I had looked upon it with such interest from a distance. I was so involved with the Arab, that I couldn't even give it a glance. It appeared that he had no thought of subsiding until I gave him a tip. I remembered then that I had almost a dollar in silver. So I finally let go of the saddle, opened my purse, trying not to slide off the camel, and began to rummage through my purse's miscellaneous contents for the money. The longer I hunted, the madder the driver got. It was very hot, and by now my head was pounding with tension. With relief I finally located the silver half dollar and quarter and gave it to the driver. Whereupon, to my amazement, he spat and threw it on the sand, saying, "This is nothing, it is nothing, it will not do!" I could have howled! But I composed myself and pulled out my purse, which was empty, and showed it to him. He gave a mean laugh and said, "You are rich American, you have lots of money." So I started all over again telling him that my husband had the money and would reward him generously if he would just continue the ride and leave me in peace. I told him he was being unkind and spoiling my ride. I should have been wise enough to appeal to the reputed Moslem courtesy to strangers. But at that moment I guess I doubted that such a virtue existed. I was hot, hurt, embarrassed, frustrated and several other things, and if I could have gotten off that camel I would have—pronto! (The Bible puts it so simply, "Rebecca lighted off her camel." Well, maybe *she* knew how.)

The driver started the camel up and we traveled a short distance in silence. Then he began all over again, asking me repeatedly where my husband was and how much he would give him. About that time I surely wished *I* knew where Dwight was! At least I knew very well what I would give *him*, if I found him—a piece of my mind. I began inwardly to blame him for everything that had happened. I also began to get scared. I couldn't see the Sphinx or any of the other pyramids, nor could I see the bus or our crowd. Just where were we? To make it worse, the driver stopped the camel again, and began haggling more vehemently. He said we would go no further until I paid him. I think we must have been between two little knolls which screened us—and our view from everyone around us. But it actually seemed to me that we were lost in the middle of the Sahara desert. He kept looking at my purse, and I sensed that he was sure that it held plenty of money—which it did, in a form I could not use. I also was wearing some jewelry. A horrible thought enflamed my mind—perhaps he would think I was worth robbing! Apparently no one was concerned about my being separated from the party; I was abandoned and in real danger. It might be hours before I was missed! I noticed that the driver had a dagger or knife in his belt, and my imagination took full advantage of the situation. How strange it all seemed! The night before I had been in an agony of fear while taking a boat ride on the Nile. Now here I was in another agony, taking a camel ride in the desert. Why didn't I have sense enough to stay on my own two feet!

I doubted that it would do any good to scream, and I realized that I had reached the kind of extremity that calls for Divine intervention. I had absorbed too much sunshine, and this, coupled with the nerve strain, made me feel faint and dizzy. I didn't like the thought of fainting and toppling off that camel. So I truly looked to the Lord interiorly with all my might. I discerned then that Satan was tormenting me in this man, and I recalled that we should never show fear in the face of an enemy. So I began to hum and smile and act as though sitting on a camel in the middle of the desert was just what I had come to Egypt to do, and I could enjoy it for hours. This had just the right effect, for the Arab muttered to himself and the camel, and began to lead him down and around a little hill, where, lo and behold, the Sphinkus crouched in disdainful indifference to my misery! Off to one side I recognized our bus. At least it hadn't left without me!

I knew that Dwight must be somewhere in the crowd milling around the Sphinx, and I felt ashamed of myself for having been so upset with him. No doubt he was very worried about me by now, and was wondering where I was. He would be so relieved when he saw me coming! And I would be relieved to see him, get some money, and get off this camel. The nearer we got to the crowd, the harder I searched for him, but I could not recognize anyone—let alone my husband! It took me a few minutes to realize that he just wasn't there! Meanwhile the little Arab was annoyed that I couldn't find him, and his annoyance, coupled with mine, made a very big annoyance indeed. My sweet forgiving attitude toward Dwight, suddenly capitulated to a feeling of outrage.

It was unbelievable that, having left so much earlier, Dwight had failed to arrive at this appointed meeting place. It was equally incredible that he had gone off somewhere else and left me stranded. But such was the case. And now I was in an even worse position. The Arab plainly did not believe that I had a husband! And he still had no intention of letting me off that camel. An Egyptian photographer appeared about then, and insisted on having me pose by the Sphinx for a picture. I almost angrily refused. I told him I was waiting for my husband. He and the driver had an animated Arabic conversation, and he stalked off saying, "You lie, you haven't got a husband." Well, of course, I am not used to being called a liar, and I began to feel that maybe he was right and I really didn't have one. When I had almost reached the point of bursting into tears, I saw Dwight coming at a galloping pace across a little knoll. He was running the camel all by himself, and obviously having a wonderful time of it. Relief and anger mingled in my heart, as he finally recognized me and rode over. Apparently he hadn't even known that I was among the missing!

Before I could say a word, he started a long recital about *his* difficulties with his camel driver! I interrupted and told him that I wanted him to pay my driver, so I could get off. Whereupon he told me that he had no intention of paying him. Our rides were already paid for, he said, and they were just trying to "gyp" us, etc. This was just too much! I became really indignant and told him that I would have to insist that he paid him, since I had promised the tip. One word led to another. And it was anything but sweet and Christlike, the way we sat on camels and argued. In the end, of course, he gave him the money—but only because he realized that I was on the verge of collapse. Whereupon my driver became his most charming self again, and offered to let me pose on the camel for some pictures with my husband. One of these pictures turned out to be quite remarkable, in spite of our upset state.

Then, at last, I "lighted off the camel"—and found out that I could hardly walk! I had supposed that Dwight and I would still have a little time to explore around the pyramids, which were nearby, and go into the region of the "City of the Dead"—a massive collection of tombs and monuments which has been excavated close to the Sphinx. But I found that he had no such idea. He had ridden off and left his driver stranded at a considerable distance away. Dwight was disgusted with him because of the way he had needled him, even after he had promised a tip, and I had never seen Dwight like he was that day. But the Egyptians are enough to get under anyone's skin, it seems. I was sick at heart to realize that we too had joined in the chorus of fussing, haggling and carnal-acting Christians!

I repented, of course, and tried to commune with the Lord while I went exploring alone. But I could hardly walk or get my breath. So I finally gave up and went to the bus. I flatly refused to get out and pose for any further pictures when Dwight finally joined me. I was sick at heart. To come thousands of miles to visit the fabulous pyramids, and then to spend the time fussing with a camel driver and my husband, seemed overwhelmingly defeating. I tried to tell Dwight that no matter what the cost, we should have paid and not fussed. But he didn't agree. Everyone on the bus was in the same mood, vying with one another as they related their feuds with the drivers. There was much talk of getting out of Egypt and staying out—it's just no place for a Christian! I couldn't help but note that our guide and bus driver took all this in and exchanged some very amused glances. One would think that after having such an experience, I would not feel a further interest in the pyramids. But, surprisingly, such has not been the case.

Although at the time I felt that never, never did I want to return to the pyramids, nor to Egypt, I was amazed to have a strong drawing to them that very night! (Though I could not follow it.) Since returning home they have become increasingly real and interesting to me. As I read about them, or think about them, I travel there in the Spirit or in my mind, and I find much of absorbing interest. I also have an interior witness that I shall see them again in person. As the Arabs say, "*Ahab, es-shoof El Ahram.*" (We are eager to see the pyramids.)

Isaiah prophesied: "In that day shall there be an altar to the Lord in the midst of the land of Egypt, and a pillar at the border thereof to the Lord." Is it purely coincidental that the Great Pyramid stands exactly on the middle dividing line of Egypt, as well as on the middle dividing line of the whole world? This is at natural zero. It's four slanted sides form the four points of the compass. It has ventilated air passages, though none of the other pyramids have such. And in it are to be found the famous "King's Chamber," the "Queen's Chamber," and various passageways. Some have believed that this pyramid was not intended for a tomb, but a temple, or an altar. And it is said that there is a tradition that it was built between four and five thousand years ago in honor of the True and Living God. Originally it was covered with white, polished limestone, which reflected the sun so gloriously that it was called "The Light."

The Great Sphinx is said to be the oldest image of man, or of God in the form of man. It is 75 feet high and 164 feet long. Its body and head are carved of natural rock, and it faces east, in front of the pyramid of Chephren. Seven different times it has been partially buried by sand, and has been excavated out of it. The most recent excavation uncovered the granite stele of Thotmes IV (1430 B.C.), which was once covered with a small temple. Only the altar of this remains. The four paws were constructed of brick, and appear like those of a lion. It also has a breast formation like a woman's. Unfortunately this monument is marred. The great headdress and beard, small parts of which are preserved in the British Museum, are missing. Moslems, or others, have shot at its nose, and Christians have hacked away at its body. But it has managed to survive the rise and fall of Egypt, Persia, Greece and Rome. And it could possibly outlive even our nation, in event of nuclear war.

In the City of the Dead, close by, it is said that strange rituals are performed at night. I was surprised to learn that the present-day Egyptians believe in some sort of communion with the dead, and that witchcraft customs, prevalent in other parts of Africa, are also practiced here. It is no wonder that we felt a little like we were in the nether regions as we explored this territory! But I hope someday to go again, in the Spirit of true praise and victory, and kneel at the little altar between the paws of the Sphinx and call upon YHVH, the True and Living God, who shall yet be exalted and worshiped in the land of Egypt.

THE GARDEN OF ALLAH

Among the few souvenirs I had purchased in Egypt was a jewelry set, fashioned of gold, copper and turquoise-colored metal. A persistent street peddler had almost forced me to buy it, and I had felt a sudden urge to do so. The necklace was made in the form of a collar or yoke, and when I had touched it, I had felt a strange sense of identity with it, though I didn't understand why. I had quickly dropped it, along with the bracelet and earrings, into my bag, and forgotten them. Imagine my surprise, then, when later I found several similar ancient necklaces on display in the Museum! In fact, in the Tutankhamen display, the queen's necklace was almost identical! It had given me a sense of awe to realize that the type of jewelry worn thousands of years ago in Egypt was still being worn today. Of course in ancient days it had been made of fine gold and precious stones. But the basic design and illusion of grandeur were still present in the cheap modern replica.

Riding back to the hotel from our visit to the pyramids, I was suddenly reminded of this necklace. Involuntarily I reached up to my throat, as though to touch it, and was surprised to find that I was not wearing it. It seemed imperative that I do so! At the same moment I struggled to say in Arabic, "I am an Egyptian." But I could not locate the right words in my limited vocabulary, so I kept inwardly repeating this declaration to myself, and to the Lord, in English. I had the sensation of being transported back to that Egypt of the past, and while I was in this entranced state everything going on around me became dim. I was gloriously detached, and my ears gave no heed to what was being said in the bevy of words being exchanged.

Yet, suddenly, one phrase penetrated my mind! I heard it distinctly, and it wafted throughout my being, like music played in the open air . . . evocative . . . nostalgic! The simple phrase was, "The Garden of Allah." And it was spoken by the warm, vibrant Arab voice of our guide, which for the moment, became to me the voice of the Lord. I felt an impelling desire to walk with Jesus in The Garden of Allah.

And the image that came into my mind was of the enchanting outdoor cafe, on the banks of the Nile, which I had seen the previous evening from the faluca. (It seemed, however, that it had been weeks ago, so much had happened in between!) During the remainder of the ride to the hotel, I sat as one rapt, held in a heavenly embrace with our Beloved, unmindful of all else. Had there been time and opportunity for prolonged meditation, I am sure that Spirit would have beautifully connected this transport with past spiritual experiences. But all too soon we were at the hotel door, and I was snatched out of my dream and back to rugged reality.

It is really amazing to me that now, almost two years after that eventful day in Cairo—as I have been meditating and waiting upon the blessed Holy Spirit—He not only has revived my memory of this experience, but has also interpreted it to me, tying it in with past spiritual experiences, and illuminating it with the glory and wisdom which only He can impart. Isn't He a wonder! It is said that the words of our mothers, spoken in our early childhood, lie very deep in our subconscious, and color all our thoughts. In my case, I know this is true, for I was strongly attached to my adorable mother, and have been aware of her influence throughout all my life, even though the Lord took her when I was only nineteen. She used to have many playful sayings, and I recall that when she would be tired and not want to work, she would say in a whimsical tone, "I guess I must have been an Egyptian Princess, in my previous incarnation, and never had to do any work, so now I want to play the lady."

Of course, my mother did not believe in reincarnation, nor do I. It was all in fun. And when I would be naughty and haughty, she would say, "Maybe you were an Egyptian Princess too, and always had your way. But now you are Frances, and you must obey." So she would not let me play long at "Egyptian Princess." Even though we do not, as grownups, play princess, nor do we go through the process of bodily reincarnations, praise God! (Wouldn't that be a dreadful thing!) We do, by the Spirit, experience many little journeys into the past, as well as into the future. In Christ we become momentarily many things—as St. Paul said—and temporarily taste of other times and races and states of existence. Wasn't it dearer of the Holy Spirit, my Heavenly Mother, to "play" Egyptian Princess with me too! I could even seem to see my strange garb that day, as well as the jewelry, and could smell a faint oriental perfume about me.

Recently, as I was waiting on Him, He reminded me also that, when I was about twelve years old, my mother decided that I needed to take some dancing lessons, so that I would learn to be graceful. A nice little teacher opened up a studio near us, and offered simple ballet lessons most reasonably—in classwork, of course. She told my mother that she would also teach us the authentic dances of other nations, and that we would give demonstrations of these, wearing appropriate costumes. Did I love that! In due time we learned the Highland Fling, and wore homemade kilt skirts; we learned La Paloma (The Dove) and used castanets and a frilly Spanish dress, among others. But the ballet that impressed me the most was ancient Egyptian! To learn the steps was difficult indeed, for it was a majestic and strictly disciplined dance. The movements of the hands must be exact, and are stiff and angular, as are the leg movements also. It in no way resembles the wild Arabic type of dance, or the exuberant Israeli *horah*! Our costumes were impressive too! I can still see my long gold-colored "shift" with its beautiful jeweled apron-like girdle, and my jeweled headdress. (My mother spent many an hour sewing on those jewels!) When I wore it I felt like a princess indeed. And the music to which we danced, played on an old cranked-up phonograph, was Arabic and exotic to my ears, if not authentic ancient Egyptian. As a closing number, my teacher asked me to sing a very nice solo, "Allah's Holiday." I learned this readily, for it appealed to me in a special way, even though I did not understand it.

Now you may wonder just what all this has to do with that day in Cairo and my sense of enchantment. So I will explain that when I had visited the Museums that morning, I had seen pictures of women wearing just such a costume as I had worn, doing dances which displayed the same positions and suggested the same movements which I had danced so many years ago. Yet these pictures were taken from tombs, where they had been painted hundreds, yes, thousands of years ago. (The colors were still bright!) This was very awesome to me. When one moves one's body in a dance of another race, there is usually a sense of identity with that nationality, just as learning to speak their tongue, or sing their songs, brings about a type of communion. To have thus danced with the dead, the ancient, the glorious ones of Egypt's causes me to catch my breath, while tears come to my eyes. I am sure you will share this sense of wonder with me, for truly it was the Spirit who was preparing me, even at that time to learn to love Egypt and its people.

I am enthralled by the folkdances of all races, just as I believe most of you are. The Spirit has taught us that rhythmic bodily movements are a form of poetry, expression and joy. And of course the highest form of dancing is that of worship and praise. "Every little movement has a meaning all its own."

Isn't it significant that just as each race has its own language, it has also its own music, art and dancing? That Egyptian dance most beautifully expressed to me a concept of what Ancient Egypt was really like. Yet I had entirely forgotten it until the Holy Spirit reminded me about it a few days ago! (Incidentally, my dancing career ended abruptly, when my mother decided that singing would be safer and more suitable for me, particularly after I was baptized into the Baptist Church, at the age of 14: and I heartedly agreed.)

The song I sang was another matter entirely! I never forgot it. Of course I realize now that "Allah's Holiday" could not have any possible connection with ancient Egypt, since Islam came to the land long after the time of Christ. In the old days they worshiped many gods, as we know, and their gods would be as abominable to us as to a modern Egyptian—for Islam is a great foe of idolatry. Allah, of course, is the one truly living God—the same God that the Jews and Christians worship. The Jews rejected Christ before He died, and the Moslems rejected Him as Savior after He died—but both worship the Father, the God of Abraham. However, in Ancient Egypt, from time to time, *the worship of the One True God was known and followed by a few*, until it was finally banished by the greedy heathen priests. Hence, my dancing teacher thought it fitting to close her ballet with some reference to God. The melody of "Allah's Holiday" is intriguing. And during the early days of our introduction to the Feasts of the Lord, the Spirit revived this song in me, and I sang it to the Lord. (I was glad that we had learned even then that Allah, Alah, and Alaha are all forms of the name of God, the Covenant-Keeping One!) Some of the words still come to me:

> "Sound of silver cymbal,
> Tambourine and timbral,
> Struck by fingers nimble,
> To some sweet lay ...
> Lovely forms are swaying;
> Raven tresses straying,
> 'Neath cool fountains' spray.
> Golden bells are tinkling,
> Breath of rose-heart sprinkling ...
> Allah's Holiday!
> Ah, could it but last,
> Could it last alway!"

The Spirit changed it to Alah's Holy Day—as I sang it over and over to the Lord. And of late it has haunted me at times again.

It has taken me a long time to lead you back to that eloquent phrase I heard on the bus—"The Garden of Allah." But the Spirit has a way of compressing or extending time in such a way that much can happen in moments—and this is what He did that day. In the natural I could have sat there crushed by the camel calamity, hurt by Dwight's seeming negligence, grieved and disheartened about my own failure to stay in the Spirit, and pitying myself because of my increasing "little miseries"—as the nuns call them. My throat was sore, my head ached, I had frequent sharp intestinal pains, one foot was very painful to step on, etc. It is all the more a joy and a wonder to me that even though the Holy Spirit doesn't always take the miseries out of our bodies, He does take us out of the miseries, by lifting us into the Spirit.

I knew by now that I was to keep some sort of tryst with the Lord, and my heart yearned toward the river and the garden I had seen. But I was troubled about how I could arrange to keep this rendezvous. I pondered this while we waited patiently in line to get our key and hurried to our room. My first step was to find that necklace and put it on! When we reached our door, however, we had a little shock. It was standing wide open and our genial Arab boy was stationed in the middle of it, holding some more disreputable looking towels—ostensibly our change of linen!

He bowed so graciously at our approach, that I wondered if he realized that I was indeed a full-blown Egyptian Princess. It would seem so! (Or was it only the tip that he had in mind?) I felt a fear about all our belongings, and as soon as we could "*shookran*" (thank) and bow him out, I rushed to my unlocked bag and made a search. About the first thing I picked up was the precious necklace. Everything seemed to be intact, but we checked through our closets too, having the uneasy feeling that he had probably preceded us in the checkup.

Then I put the necklace on, inwardly rejoicing. It seemed to instantly heighten my awareness of being an Egyptian. But I noted Dwight's disapproval, out of one eye, and ignored it, busying myself with collecting some laundry to do. Scarcely five minutes passed until our Arab boy reappeared, suddenly, and without knocking. There he stood in the middle of our room, smiling and repeating, "halo" and "please" over and over. He was getting just too companionable for words! And we were fresh out of words—that is, words he could understand. In the moments that followed we had a pertinent demonstration of how frustrating the language barrier can be, when communication is urgent. He obviously wanted to convey some message to us, but what he was trying to say we could not grasp. He was more resourceful than we, so soon he resorted to the universal sign language. First he held out his hand to us, showing us that he wanted us to give him something, then he bowed to indicate his gratitude for the forthcoming something or other. Next he made a gesture as though in pain, and closed his eyes, holding his hand to his head. Now anyone should have known what he was saying, but I guess we were still too surprised to react normally. So he began the word struggle again, finally getting out this astounding question, "You—doc-TOOR?" meanwhile bowing at Dwight and pointing to our array of vitamins, proteins tablets, medication and cosmetics—all neatly arranged on our one shelf. (Come to think of it, it did look a little like a private drugstore!) We finally caught on! He had a bad headache, and thought Dwight must be a doctor, in view of all those pills. Dwight sprang into action and said, "Sure, sure!" He got him two aspirin and patted him affectionately, while the lad bowed and exuded gratitude, almost to the point of kissing Dwight's hand. I had a curious feeling, however, that maybe he had already sampled our various assortment—and I hoped there would be no drastic consequences.

About this time Dwight remembered that he had to see one of the brothers—so out he went. And I decided it was time to sally forth, wearing my necklace and sweetest smile, to see what could be done about getting our passports, some money, and especially, our mail. Now that I was the real Egyptian, I was sure things would be different. But my newborn confidence was in for a tussle. In the first place, the key would not lock the door of our room, and I was not about to go and leave it open. Obviously we have been given the wrong key. Now I simply *had* to be understood. So I hastened to the desk clerk. After greeting him in Arabic, I showed him the key, shook my head and then told him in English that it would not work. He agreed, and shook his head along with me, and then went to answer the phone. I waited while he carried on an explosive conversation in Arabic. Meanwhile two busboys were watching me with candid Arab amusement, and I knew it was the necklace that had caught their eye. When I greeted them they responded and pointed to it, making it plain that they were pleased. But that was just as far as it went. I could not make them understand about the key either. The clerk had now disappeared, so I went on a further search. Down the hall was a little office, and in it I found two girls typing. So I went through the greetings again, and then showed them the key also. But they only shook their heads too. Ah me! Then one pointed to my necklace and commented about it in French to the other one. Of course! I should have remembered that French was widely spoken in Egypt at one time, and is still used more than English. I started all over, this time in French: *"Bonjour! S'il vous plait, je desire a dire avec vous."* *"Mais certainement, Madamé, avec pleasure."* Ah oui! Now I was off to a good start, and with the Lord's help, we carried on an animated conversation. She admired the necklace Madamé was wearing, and was glad she was learning Arabic. She wanted to come to America and write for the "seen-e-ma"—which I finally figured out was *cinema*—movies. I announced I was *"Ana El Masr sitt"* (Egyptian lady, I think.) And all was most convivial between us. She readily understood about the *"clé"* and the *"porte"* and went to the desk clerk and got me the right one. She also got him to agree to cash some travelers checks, so we could have some money. But as for the mail and the passports, I would have to see our travel agent, and he was apparently hiding out for the duration.

I was as happy as a lark when I returned to my room. My Egyptian Princess role was becoming more enjoyable by the moment. I doubt that any of them ever did their own laundry, but I enjoyed doing mine and hanging it out on our little balcony. As I stepped out to do so, a curious thing happened. From below came the sound of trumpets and auto horns, and I could see that a parade was in progress. How impressive! And I could review it all, just as Cleopatra might have done, from my balcony. Since Sukarno was visiting, I realized that doubtless it was in his honor. And soon, sure enough, here came Nasser and Sukarno, riding in an open car! I felt such a love for each of them, knowing that their nations are part of our "family." And yet the incongruity of the situation filled me with laughter. Here was I, hanging out washing, in full view of these leaders, and of the ancient pyramids, and feeling all of you present with me, playing at being Egyptian Princesses too. A rare and unforgettable moment!

After a rest period, I mentioned to Dwight that I would love to walk to the Nile that night. I stressed *walk*—since I had decided to stay on my own two feet, as you will remember. He readily agreed to go, to my surprise, and we both felt that we would enjoy being alone together and away from all the group.

The atmosphere at dinner was depressing—most of those present were still rehashing their exasperating experiences with the camel drivers, guides, street peddlers and beggars. Others were talking about how ill some had become, and all the dangers to be encountered in Cairo. It was agreed that no one should go out on the streets at night, unless in the group etc. But in spite of all this, Dwight and I ate in peace and then set out upon our little quest. When he told our genial "watchman"—who had reappeared in the lobby after a short absence—that we were going to walk to the Nile, he seemed flabbergasted. He advised us to take a taxi, by all means, and said it was a long distance. But I was intent upon walking, in spite of my very sore foot. And I know it was the Spirit who impelled me.

He led us to take just the right street. We had a choice of at least three that led in that direction; but unerringly we were moved toward the very one Nasser had ridden down that afternoon. It was a wide boulevard and though the traffic was heavy, the sidewalk was strangely deserted. Policeman, dressed in white uniforms, were stationed at intervals all along the way, and I'm sure that because of their presence the crowd felt uneasy about using it. They appeared rather ominous, in spite of their short stature and badly wrinkled clothes. And the guns they carry are large. I could not help but contrast them with our rather genial officers of the law, with whom we are urged to be friendly. Not so in Egypt!

The natives may choose to avoid them, but we were thankful for them as we walked in a leisurely manner toward the river. In my eyes they appeared like guardian angels, and I felt that I was taking a royal promenade. There is something about walking—about setting one's feet down and taking time to look at things along the way—that brings a close identity with a region. I forgot my painful heel and all my other ailments, in the joy of walking on what seemed to me to be the King's Highway, through Cairo to the banks of the Nile. No one tried to sell us anything, no one jostled us, and few even stared. Our only human contact was with a fine looking Japanese, with whom Dwight enjoyed a short conversation in his language.

The distance to the river was really long—though in my thoughts it had seemed to be but a few blocks. We passed many fine buildings, including the Palace where Nasser was entertaining Sukarno, after dining with him at the Nile Hilton. It was closely guarded, but we stopped long enough to feel a contact with these men. When we finally reached the beautiful banks of the river, we sat and rested on a bench, opposite the Nile Hilton, enjoying the beauty and elegance on every side. I waited until we were rested somewhat, and then said that I would enjoy walking across the famous and most beautiful bridge to the island. From where we sat it appeared like a starlit archway into some mystic realm. I knew, of course, that it led to my own personal "Garden of Allah." And I was eager to follow it. Dwight, however, was very tired, and insisted that if we went that far, we would have to return to the hotel by taxi. I was pretty tired too, so I relented and agreed to risk another taxi ride through the streets of Cairo.

Crossing that bridge was simply celestial, in spite of the rush of traffic beside us. The river flowing below was seemingly filled with floating stars, because of the reflection of many lights. On the island the great tower of Cairo stood in majesty, lifting its huge torch high overall. Ancient Egypt, modern Egypt, and the Egypt yet to come—in the Kingdom Age—all seemed to be symbolized in the scene. The Nile, of course, was ages old; the beautiful bridge and huge buildings were modern architectural masterpieces; but the tower seemed to be something exalted—symbolic of the day when Christ reclaims this land and is loved and served by its people.

It was not difficult to steer Dwight to the café and garden, once we had crossed the bridge. Fortunately it was close by! And he was ready for a nice cup of tea. I felt the Lord desired to have a love communion with me there, with a glass of wine. But knowing how opposed Dwight is to wine, I was hesitant. He seemed to sense my feelings, for when I mentioned it, he readily agreed and ordered some for me. We sat there in silence for at least a half-hour, slowly sipping—he, the tea, I, the wine. On T.V. close by, a girl was singing Arab songs with gusto. I didn't understand the words, but the music was most fitting for the Garden of Allah, and someway it all seemed to create just the right atmosphere for a most precious interior communion with the Lord. I felt that night that Jesus was an Egyptian, the gracious Pharaoh of the Land. And I longed for the day when He shall take His rightful place in the hearts of these, my people.

The charm of that evening in Cairo is one of my sweetest and most cherished memories of the trip. Coming at the close of a most frustrating and disappointing day, it was all the more appreciated and welcome. Truly our God can set a table for us anywhere—in the desert, by the sea, in a great metropolis—just as He set the table for Israel in the wilderness. "Thou preparest a table before me, Thou anointest my head with oil, my cup runneth over." This He most surely did for me on that memorable night, in the Garden of Allah, on the banks of the Ancient Nile.

En Neel, En Neel,
The pyramids in majesty
En Neel, Gameel!
Guard you silently,
Here in Egypt's ancient land,
And the sphinx wears a cryptic smile
I walk upon your golden sand
For she knows your secrets, my Nile.
And in a mystic daze
En Neel, En Neel,
Upon your beauty I gaze
En Neel, Gameel,
River of life and love,
All the glory of Egypt's story
Like to the one above,
Is reflected in thee, I feel,
En Neel, En Neel, Gameel!

MEETING THE LORD
AT THE TOMBS OF
THE SACRED BULLS

O f all the places in Egypt where I might have expected to meet the Lord, the Tombs of the Sacred Bulls, at Memphis, seemed one of the most unlikely and undesirable. Yet there was a rendezvous awaiting me there, both with the Lord and with semi-disaster! I have always had a special antipathy about those Apis bulls, which the Egyptians adored and entombed in regal splendor almost equaling the fantastic manner in which they buried their kings. Not only in life were those chosen bulls fed and cared for with utmost delicacy and reverence, but, even in death, each bull was adorned like a bride and placed in a specially prepared tomb, lying in a sarcophagus of gold and precious jewels! I suppose I resent all this deeply because of the golden calf which was made by Israel in the wilderness, in imitation of this vile bull worship. So, of course, I wouldn't have gone one step of the way toward their tombs, had not they been located in the ruins of ancient Memphis, a city I wanted to visit because it has always fascinated me.

There were dozens of things in Egypt I had hoped to see, things vitally connected with the Scriptures or the history of the Church. And this was only to be expected, since I had read some of the colorful and inviting folders which the government publishes to lure American tourists. According to these, Egypt is a land abounding with places of interest to Christians. From these pamphlets, one gets the impression that Egypt was a stronghold of the Bible and of early Christianity, and that it is an outstandingly Christian nation even today. The first part of this concept is true. We all know that Israel was hidden in Egypt for hundreds of years, and that many Jews lived in Egypt both before and after the time of Christ. Alexandria was a world center of culture. And the famous Septuagint translation of the Old Testament was made there, along about 300 B.C. In the early days of Christianity, Luke, Mark and other disciples and evangelists lived and worked and wrote in Egypt. Many of the ancient biblical manuscripts that have been discovered were found in Egypt, safely preserved in its hot and perpetually dry climate. One of the most important of these is the *Codex Sinaiticus*—discovered in a monastery at Sinai.

Likewise, the Coptic Church of Egypt and Ethiopia have preserved many of the traditions and practices of the primitive Christian Church, and to this very hour are perpetuating them in the earth. For hundreds of years they were completely separated from all other branches of the Christian Church—due to a strong breach—hence, they are free from outside influences. There are many things to see among the Copts, including a Museum of early Christian writings and objects of interest. However, when it comes to visiting or even finding these churches, the tourist is usually completely frustrated. For the regrettable fact is that although the government is very interested in Christianity, when it comes to travel propaganda, its interest ends abruptly when you set foot in Egypt. All tourists guides are licensed by the government and directly responsible to them for their conduct. Therefore we must assume that they reflect the attitude of the government, *which is centered in Nasser.* And our experience was that the guides had no intention in the world of taking any of us to these places. We were met with subtle evasive measures, or with feigned ignorance. I have learned that this is the common experience of all tourists in Cairo, and that to get to the old section of Cairo, where the Copts live and where there are really ancient churches and things of special interest to Christians, is a virtual impossibility, unless one has a private guide, and is prepared to take some risks. For centuries after Islam was established, the Coptic Christians were most cruelly persecuted. Therefore, they hid their churches by most ingenious measures—underground often, or disguised by false fronts, etc. And they are still hidden, though the worst of the persecution has ceased. At least they are hidden from the tourists.

It was most baffling to be so near famous places of biblical interest, and not be able to visit one of them, nor to locate any modern evangelical churches in Cairo—though there are some there. (Near the pyramids there is a fine Pentecostal Church in which Maude preached and where we could have had fellowship, but we knew nothing about it then.) Just a few miles northeast of Cairo is a place called Mataria (Matar-EEH) where, it is believed, Joseph and Mary dwelt for at least part of their sojourn in Egypt. This is located near ancient Heliopolis or On, where the Old

Testament Joseph lived, loved, and served God. Moses, also, is said to have lived and received his education as a prince in this area. A red granite obelisk, 66 feet high, is still standing there, which was supposedly built in the time of Abraham. It is the last of several that stood in the front of the Temple of the Sun. It was doubtless often seen by Joseph, who married Asenath, a daughter of the priest of On. Moses too must have looked upon it. And then, centuries later, Joseph, Mary and Jesus lived there. (It is of great interest to us that three other *obelisks* from this ancient temple are now located in New York, Rome and London.)

The Coptic Christians believe that the Holy Family rested in a garden in Mataria and that as the little feet of Jesus touched the earth two springs of sweet, clear water sprang up. The springs are still there! All other water in this community is brackish! This garden has been renowned for centuries. Up until the time of the Middle Ages, balsam trees flourished there—though they were nowhere else to be found in Egypt. (Cleopatra had planted them from cuttings she brought from Jericho.) Unfortunately these died out in the seventeenth century. French Jesuits have built a little church next to the garden of herbs which marks this spot. It is called The Church of The Holy Family. Its walls have frescoes which preserve the story of their entry into Egypt. These show that when the Holy Child entered Heliopolis, the noise of a mighty wind was heard, and idols fell from their pedestals, as the earth shook. (As we recently read in the New Testament Apocrypha.) Now who wouldn't want to take a little journey to Mataria to see these things and walk in the region where the holy ones of old once walked! But we could not find any way to manage such a pilgrimage, during our short stay.

Another place our fellow tour members discussed wanting to visit was the orphanage of dear Sister Lillian Thrasher at Assiout. But it was far up the Nile, and hours away even by plane. (She died shortly after we were there.) We talked of all these things at the tables at meal times, and as we gathered together in the buses. And it was inevitable that we felt somewhat cheated and disappointed. A number grew quite rebellious about being "dragged around to see all these Moslem and heathen things." And I am sure that if we had made out a list of all the things in Cairo we *didn't* want to see, the tombs of the sacred Apis bulls, recently excavated at Memphis, would have been near the top of the list. Yet it was just this little "treat" that our guides had cooked up for us to "enjoy" on our last day of touring in Egypt.

It's no wonder, then, that I woke up that morning feeling as heavy as though one of those bulls was sitting on my chest. I was oppressed, depressed, and just plain distressed from head to foot! All the blessing and enchantment of my little "dream sequence" in The Garden of Allah, had completely vanished, as though by a puff of smoke from the mouth of a genie. I had a sense of foreboding and confusion difficult to define in words. So, since it was still rather early, I lay still, trying to praise and commune with the Lord. But He seemed to have folded His tent and departed as silently and swiftly as an Arab. My mind began to race forward and backward. Tomorrow, I told myself, we shall fly back to Jerusalem. Oh joy! And Friday will be the beginning of our Pentecost in the Holy City! It was just incredible, and seemed almost as remote as the millennium! Even to think of it was a torment, for I was suddenly projected backward to the preceding Wednesday: the agony of the postponement of the flight, the fears, the uncertainties we all met concerning my departure—these suddenly revived and I was in a travail again—far worse than what the actual experience had been. It seemed now to have caught up with me—the *reality* of all that happened.

I tried to praise the Lord for having overcome every difficulty and safely transported us; but a greater fear attacked me—that we would be delayed in Egypt through illness or accident, and not get to Jerusalem for Pentecost, after all! It was so real and clear that I KNEW it was so. I could picture the pain of having to remain in Cairo, and the consternation of all of you, if, after all, I could not be present in Jerusalem.

The torment was just too great, so I decided to arise, try to keep praising the Lord inwardly, and ignore this attack. But I was still exhausted and dazed, and in much pain. Too much exposure to the sun had resulted in my usual reaction to such—a pounding head and trembling nerves. The pain in my foot had increased considerably and I discovered an odd sort of growth on my heel. My back and legs were in misery too, and I had a sore throat and a cold in my head. (Sore throats with me usually mean a month-long siege with a bronchial cold. So this was dismaying.) But my next test was positively nightmarish—I suddenly noticed that my right arm was broken out with a bright red rash! That did it! Now I knew the worst—I had the measles! In the excitement of the trip, I had managed to forget that I had been exposed, just before leaving. Of course! The sore throat and cold, the aches and general all-over feeling of misery—these were symptoms of measles. And if I had them, I was doomed to a hospital or a hotel room until I recovered. By that time everyone would have left Israel and the tour would be over. This was ghastly!

It was Dwight who brought me out of my state of semi-shock, by bursting into the room with news of his early morning skirmishes with the natives. I tried not to let on to him that anything was wrong, for by now I was determined to conceal the measles with calamine lotion, if possible, and tell everyone I had an allergy, (which I have.) How wildly desperate we become at times—in off-guard moments of weakness! As soon as Dwight left, I began to really call on the Lord in earnest. I admit that I was NOT willing to have the measles or any other incapacitating illness that would keep me in Egypt. I cried to Him to please, oh PLEASE, if I HAD to be sick, let it be in Israel—which suddenly became HOME to me. After a few moments, the Lord must have quieted me, and I began to realize that measles takes longer than eight days to incubate. Also it does not usually begin on one's arms. So I hunted carefully, but could not find any red spots anywhere else, and my fear began to subside. I was ashamed then, for being such a big baby. But being sick away from home has always been a big bogy to me . . . and to be sick in Egypt seemed overwhelming!

The financial aspect was equally fearsome, for if we had to stay in the hotel, or a hospital, for even two weeks, we would have nothing left on which to travel home. If we did not move when the tour moved, then we would be stranded. The fact that I had trip health insurance was not too comforting, for it might take weeks to settle that. But, of course, the worst prospect was to miss the pilgrimage to Jerusalem! By now I was sure that I did not have measles, but if not—what was it? I doused my arm with calamine lotion and hurried to dress and go to breakfast, being so relieved that I was able now to ignore all my other discomforts.

Breakfast proved somewhat doleful that morning. A number were ill, and one little elderly sister, who was often attached to me, was trying bravely to drink some coffee. She said she was still sick and very weak, and could not see how she could possibly pack and be ready to travel the next day. We both agreed that we HAD to make it, and that someway the Lord would help. I was moved to much pity for her and all the sick ones. Besides, there was still no mail! And our "detective" friend appeared somewhat grim toward us—or so I thought. (I guess he was sick too, but it made me shaky.) Why no mail, no passports, no travel agent! Maybe we were stranded and none of us would get out. All sorts of rumors were being aired. And no one, but no one, wanted to go see the tombs of the bulls.

Nevertheless, everyone who was physically able "went along for the ride." And it was indeed a hot, uninspiring affair. All the windows were wide open, and my cold kept growing worse. The one thing I really wanted was a big cold drink of water—I suppose I was feverish. Everything I looked on was most sordid and depressing. The Lord was not "playing princess" with me now! I felt like a refugee, as we left Cairo behind and turned toward the ruins of Memphis, the very ancient royal city, which lies about 12 miles south of Cairo. It is difficult now to picture how magnificent the city once was, for it was destroyed entirely by the Arabs in the seventh century. It is said that if one flies over it, one can see the remains of a vast series of pyramids and other tombs, extending more than 60 miles along the Nile! The famous step-pyramid of King Zozer, erected almost five thousand years ago, is nearby. Although the city of the living has disappeared, the city of the dead is being excavated by degrees, and new discoveries are being frequently made. The dry sand is a perfect preserver, and many of the tombs are in excellent condition. The treasures were stolen centuries ago, but the stones and decorated walls remain.

However, I was not thinking of the Egypt of the *past* but of the impoverished Egypt of the *present*, as we jogged along. Once outside of Cairo, the little mud houses of the *fellahin* increased. Built close to canals, running off the Nile, they looked like small mounds of earth, rather than houses. They were not even as large as our Arizona Indian hogans! We were told that in these low, tiny huts, the entire family live—including their animals. Incredible! They wash their clothes and bathe in the dirty canals, and drink from them too, along with their animals. (Some of them had a water buffalo to help plow, and a few had goats.) With the ancient *Shadoof*, consisting of a bucket fastened to a pole, they draw water to irrigate their crops.

The women all wore the black *maleya*, which resembles a nun's garment, except that the veil is long and more or less conceals the face at all times. These Egyptian farming women are said to be old at twenty-five, and they all appeared most depressed, sad, thin and worn. To me they appeared like black ghosts out of Egypt's glorious past. Even the little children looked much too thin, and were clad in ragged clothes. They had none of the exuberant nature of the Jordanian Arab children, who appeared happy and plump, even though very poor. I was sick with pity for the women and children. And seeing how they lived, I could not wonder that in some areas of Egypt ninety percent of the inhabitants suffer with snail fever. It flourishes in these canals, and is a most terrible wasting disease, for which no cure has been found. (I'm glad I didn't know as much about it then as I do now, or I would have felt even worse—and it is easily contracted by travelers, sad to say.)

Whenever the bus stopped, usually to take pictures, the nearest family rushed to the road and started begging. The children were simply heartbreaking! But they will not pick up our silver, for some funny reason, and one would soon run out of dollars. So one just has to smile and try not to mind—but it grew increasingly difficult. I was increasingly depressed at the way our Tour members were taking pictures of the shocking things, and laughing and joking about them, in a very superior manner.

At one place we passed a sort of outdoor meat market, which was so revolting a sight and smell as to make all of us sick and disgusted. The flies were there in myriads. And we learned, to our dismay, that it was from this market that most of the meat used in Cairo is purchased. This was enough to put one on a fast! (We always wondered what we were eating. The meat was all tender but we suspected that some of it was camel or goat.) Our guide seemed to be highly amused at our consternation about this disgusting meat market. The Arabs do not share our fear of germs, flies and dirt! Likewise, he was not disturbed about the miserable living conditions of the *fellahin*. He assured us that the people were very happy and contented, and that they enjoyed living this way. Then he began again to sing the praises of Nasser, recounting all the improvements he has made in the redistribution of the land and economic improvement. In Cairo, radios blasted from almost every dwelling place. And although only the better class can afford a T.V. in the home, the government installs public sets in the parks and squares, so all the people can watch. The Arab does not covet privacy the way we do, and enjoys mingling with others in every phase of his life. In the cities the streets are crowded all the time, even at night. In the villages the people congregate in the markets or some public area. The *fellahin* man works from dawn to dark, and then sleeps on the dirt floor of his hut until another sunrise. But he likes it that way—or so we were told!

We finally came to a stop at a pretty little oasis-like place, where a mammoth statue of Ramses II has been excavated. It is so large that they have left it lying down on its back, as they found it, and they are now building a Museum to house this colossal figure. There was a small sphinx here too, and other interesting antiquities. Young boys were digging in the various piles of sand and now and then discovering tiny pieces of stone and such that were centuries old. These they sell to the tourists for a pittance. I finally gave in to one and bought a scarab, a carved stone and a tiny figure of Anubis—all for a dollar! While we waited, another *walad* (boy) climbed a tall palm tree and hung in precarious fashion, posing for the benefit of those taking snapshots and movies. He was most amusing and obliging. Then, after they had finished, he slid down and held out his hands to them for a little gift. To my surprise, not one of them gave him anything! I was so shamed of the way they acted that I could've cried. They just laughed and hurried onto the bus, leaving him standing with the most troubled look imaginable. He was so skinny and ragged that I was sick. I can still see his hurt eyes, and it pains me. I should have taken the initiative and given him the dollar, even if I hadn't shot his picture. But I hesitated too long, and when I looked for him, he had disappeared.

Several of our tour became upset with the souvenir peddlers, and spoke to them harshly. And I heard one pious minister threaten to punch one of them in the nose if he didn't leave him alone. I could not help but be ashamed and disturbed about the attitude of our companions. So I was near tears when we got back into the bus.

The next stop was at the tombs of the Apis bulls. What a desolate place! It reminded me of the scene from *The Ten Commandments*, where Moses was driven out into the desert wilderness to die. The wind was blowing the sand all around us. And it was very hot. We found we had to wade through the golden sand, which seemed suddenly to have lost its beauty, now that I had to meet it feet first. (Much of Egypt is golden, but the sand in some areas varies from black to snow white.) I was by now thirsty enough to drink any kind of water, but the Nile was not at hand. So I felt like a dying pilgrim indeed, plodding toward the tomb. By the time we reached the entrance and started underground, we were all thankful for the relative coolness, even though our eyes could not see anything, after the glare of the sun. Our guide led us down a long underground tunnel, dimly lighted at intervals, and the air grew heavier as we walked. Talk about being buried alive! This was it! I was reminded of the scene in the opera *Aida* where she and her lover were thus punished. I can still recall how horrible it was, and how I marveled that they each sang until they expired. I was so stifled by now that the very thought of singing was staggering. What a place to have a noonday song! I wish now that someone would go there and burst forth in praise to the Lord! But at the time I could not seem to do it, and not a word of praise or of the Lord did I hear from any of our people.

At various places along this tunnel there were openings which were the actual tombs in which a bull had been buried, these too were dimly illuminated. In some of them there were amazing, unretouched paintings on the walls, still bright and colorful after thousands of years! The scenes depicted were of Egyptian life in those far-off days, and I wondered how it could have been that less than twenty-four hours earlier I had felt identified with Egypt and all its past, present and future. Now it was all utterly foreign! Dwight was not at all happy in this underground maze, and soon restrained me from continuing further. The air was very bad, and one could easily imagine that it was filled with disease and death. I doubt that the tombs varied much anyway, and I had seen enough to know that the sarcophagi given to the bulls were as elegant as those provided for the pharaohs. It seems incredible that a race as advanced as the ancient Egyptians could have believed that their god Osiris lived on earth in these bulls (one at a time.) Not only did they worship and pray to their god in this form, but they also regarded the Apis bull as an oracle. When a question of special importance had to be settled, their rulers consulted the bull. Two special sanctuary-like chambers were provided. If the bull entered the one to the right, the answer was yes; if he chose the one on the left, the answer was no. (Or it may have been the other way around.) The priests, however, had less faith in the bull's oracular ability than the people. So they settled the issue by putting delicacies that would attract his highness into whichever chamber they wanted him to go. And then, in a great ceremony they inquired of him, and he obliged them by following his nose to the treat! Such foolishness! But, even today the Hindu regards the cow as sacred—believing that the universal mother-god spirit is in her. So millions of humans beings starve, while the sacred cow eats and does not get eaten.

As I walked down these ancient halls, I had only one thought—how do I get out of here? So I failed to recall the great personages who had walked there before me. Not only the great of Egypt came to honor the entombed bulls, but also the renowned of many nations: among these Alexander the Great! This remarkable man, who went to the Temple in Jerusalem and honored YHVH, also paid his due respects to the Apis bulls at Memphis. If only he had been truly converted to the God of Israel, how different might have been his end!

Once we reached the land of the living, I was all for collapsing in the bus and hurrying back to the hotel. But not so! Our guide insisted that now we would go to the finest tomb of them all, wherein many notable antiquities were preserved. And, of course, it was an uncomfortable distance away. My fever was mounting, my throat was parched "like a potsherd," and I felt like the Psalmist who was about to perish. Trying to walk in that sand gave me a new appreciation of the journeyings of the Children of Israel. When Dwight saw that I was on the verge of exhaustion, he suggested that we get a donkey cart, which conveniently appeared at that moment. There was apparently a little road to the tomb, and we could ride! Hurrah! I forgot my resolve to stay on my own two feet, and agreed to risk it. Surely a donkey cart couldn't be as awful an experience as a camel ride! Dwight gave the driver to understand that he would pay him his fee and a tip *only* if he kept still and did not nag us. He was to take us to the tomb and wait for us, then take us to the bus. They agreed on a price and in we climbed. A few others got in carts too, and for a few moments we all cheered up. But when we got to the tomb we had to wait, of course, for those who had walked. So we baked and stood about feeling quite miserable. Dwight refused to go in at all, but I trailed along. And it did prove to be a splendid tomb indeed, with an elegant sarcophagus and other furnishings, wall paintings, etc. But being buried alive twice in one morning was just too much! Besides, it was noon by now and we were starved, after our meager "continental" breakfast.

Outside, Dwight was haggling with the driver. Either he had not understood, or else haggling was his only fun. He insisted on being paid immediately and leaving. Dwight, of course, insisted he wait. They were both seemingly angry when I reached the cart, so we departed in a cloud of sand and ill-will. And all the way to the bus, the fight went on. Fortunately we hadn't far to go. The driver stopped at a considerable distance from the bus, and refused to budge. After more heated discussion, I urged Dwight to get out. I had spied a cold drink cart over near the bus, and I felt like a desert-dehydrated fugitive who has just glimpsed an oasis. So Dwight got out and started to walk away, as though he would not pay! Of course, the driver pursued, and I feared they were really going to come to blows. It was a most unnerving moment for me, climaxing an unnerving morning. So I decided to make my escape. Since neither of the men showed any interest in helping me alight, I crawled out on the step and prepared to jump. I have no idea what happened—but the next thing I knew I was plummeting earthward, in a most distressing fashion, definitely *not* feet first.

How one can have a whole chain of thoughts in a few seconds, is hard to explain. But I did! Three times in my life I have had very bad falls, and each time I injured the coccyx area seriously. After the last one, which occurred right after Jody was born, I was unable to sit down for a month. I had to stand or lie down, and every moment was agonizing. After the first two I was unable to walk for a short period of time. These experiences raced through my mind in vivid sequence, and a horrible fear accompanied them—if I could not walk or sit, I therefore would not be able to continue the trip. For the second time that day, I saw myself in a hospital. But about then I landed in a huge bump. I never dreamed a sand road could suddenly become as hard as that was! There was a moment of semi-shock, and then excruciating pain. Neither Dwight nor the driver saw me, but two other carts passed then and everyone began to holler. Meanwhile I had lost my breath, but was inwardly crying out to the Lord. Also I was very embarrassed—especially because my husband had not seen me and rushed to my rescue, and the other tourists were trying to attract his attention. So I began to get up, exercising all the faith I could muster, and rising in the name of the Lord. He met me in a most wonderful way, all praise to Him! And by the time Dwight finally did reach me, along with the now very apologetic driver, I knew that I could walk. But, could I sit again? This was the big question. Dwight quickly settled our bill then, and escorted me slowly and painfully over to the bus—a distance of more than a city block. There was a bench there, and I found that I could sit too!

Oh wonderful! With a cold drink in hand, I gradually revived. Everyone was fussing over me and Dwight kept getting after me for trying to get out alone. In my heart I knew that once again I had acted in my own spirit. I hadn't looked to the Lord in the cart, but had tried to run away. And I was really chastened.

By the time all our number were collected in the bus, we were a subdued and dejected lot. After all, even the most robust tourist cannot escape suffering from the lack of drinking fountains and restrooms. Added to my already long list of miseries was the pain of the recent fall—which affected me from my neck to my feet. The constant jerking of the bus aggravated all the painful places, and only by an intense effort of praise could I keep from moaning and crying. Then, suddenly, in His own wonderful manner, Our Lord not only visited me, but entered into me in a palpable way! I could feel Him within my pain-wracked body, sharing with me both my mental and physical difficulties. It did not take the pain away—He sanctified it! And I remembered how once He had whispered to me that He had sanctified His "house" (meaning us) not only by the Word and the Spirit, but also by "purging pain." I knew that once again I was cast into the fiery furnace.

Then, in a most wonderful manner, I felt His eyes in my own eyes, and His heart crowded into mine. And I began to look out of the windows upon the Egyptians in union with Him. A flood of understanding, love and compassion overflowed my heart and rose like an artesian fountain into my eyes. Tears flowed copiously, and I could not restrain them, though at first I tried. (I suppose I hated for the others to see me weeping, since they would attribute my tears to the fall.) All the way back to the hotel I was rapt in union with Christ. And as I now recall it most vividly, I realize that it was much like the night I entered Jerusalem—so strongly did the Spirit move. Jesus continued to look through my eyes into the faces of all who drew near our bus. And some way I felt that His tears, flowing through me, interceded for the souls of that region efficaciously.

I also realize clearly today, though I did not then, that it was in the place of death (the tombs), and in the power of pain, that His union with me as "an Egyptian princess" was consummated. On the preceding evening I had tasted of the beauty, joy and wonder of it. Had it ended there, it would have been just a romantic spiritual experience. I am glad that our Lord takes us on from *romance* into rugged *reality*. If

I was to be truly identified with Egypt and her people, I must share their lot: their needs, sufferings and conditions. Hence I was given a rather large dose of sickness, weakness, pain, fear, torment, hunger, thirst, darkness and the tomb. For this is most certainly the common lot of the Egyptian. In spite of the comfort of their Moslem faith, there is much spiritualism, witchcraft and other works of darkness in Egypt. It is said that they regularly visit the "cities of the dead," trying to communicate with their ancestors and lost loved ones. And Cairo itself is a place where "hell holds high festival"—according to those who know it well. The drug traffic, white slavery, murder, robbery and political intrigue all center in this strange city of between two and three million souls.

In addition to all this evil, the government continually sends forth words of hate by radio, the press and personal envoys, stirring up other nations to hatred. Hatred of Israel comes first. Hatred of King Hussein of Jordan runs a close second. Also on the list are Saudi Arabia and its king, Libya and other nearby states. The Imam of Yemen was driven out by this fomented hatred. A nation that sows such hate cannot help but reap it. It is no wonder that we all felt such evil forces there that we were eager to depart. I marveled that anyone could do fruitful Christian work in such a land! Yet, after meeting The Living God at the tombs of the dead and vanished gods of Egypt, I was willing, if He willed, to remain in the land and love these people as He loves them. And I can gladly say that of all the experiences of the trip, none was more poignant and unforgettable than my rendezvous with Jesus at the tombs of the Apis bulls.

P.S. Just two years ago today (as I am writing this), we were in Egypt, living the experience of which I have written. It is truly marvelous to me the way the Spirit helps me to relive and interpret the multitudinous and varied experiences of the trip. And I rejoice too that so many of you relive them now with me.

THE ALABASTER MOSQUE
AND THE IVORY TOWER

A mong the treasures found in the tomb of Tutankhamen were a number of articles made of oriental alabaster. These not only attracted but fascinated me—when I viewed them at the Museum—so exquisite were they in beauty and design! The delicate coloring, artistic marking, and superb translucency of this fine alabaster gave these articles such a celestial quality that they seemed to me to have come from heaven, rather than from an ancient earthly civilization. I lingered behind to admire this display, and doubtless missed part of the Museum guide's lecture. But oh, how the alabaster spoke to my heart! My love affair with alabaster began a long time ago, when I was at the impressionable age of eleven. A charming Sunday School teacher told us the story of the alabaster box of precious ointment which was lavished by Mary upon our Lord. She made the story seem most beautiful and inspiring to my young heart. And from that moment on I loved the sound of the word alabaster, and was attracted to anything made of it. It seemed most fitting that it be used for royalty and for worship. Later, during the holy bridal times of the Spirit, the breaking of the alabaster box or flask was made a very personal reality. But little did I realize that in the time of David—and even much earlier—alabaster flasks were being used in Egypt to preserve precious perfumes and ointments! It was believed that nothing served this purpose as well as alabaster. It is said that some of these, excavated after thousands of years, still retain the fragrance of the perfume they once held! Neither did I realize that the oriental alabaster used for such flasks was of a different substance than the common marble-like alabaster which is utilized today.

It was fitting that I made new discoveries about alabaster in the land of Egypt, for it was here that it was first discovered and named *Al Albastron* for the place where it was found. It was also fitting to discover that there is still much alabaster in Egypt—though of a less valuable quality—and that *Cairo has a mosque made entirely of this beautiful substance!* This startling news hit me like a bolt of lightning, as we neared our hotel, returning from the painful excursion to the tombs of the Apis bulls. I was occupied with the moving of the Spirit, and paid little heed to the running commentary being carried on by our guide. However, as we neared our hotel, he called for our attention, while he announced that after lunch we would visit the famous citadel of Cairo and the beautiful Alabaster Mosque. I could not believe that anywhere in the world there was a great building made entirely of alabaster—and I did not conceal my incredulity. So the guide proceeded to further enlighten me about it, to everyone's amusement. They, apparently, thought it most ordinary to make a great building out of precious alabaster!

I began to wonder if I was a little "touched" after all, and was reminded of how pained I had once felt about the ivory palaces of our Lord, which I first heard described by this same Sunday School teacher. Everyone else in the class had thought it most ordinary to make palaces out of ivory. But I was shocked for the thought of killing myriads of elephants to supply the tusks to make Jesus such elegant palaces had seemed cruel to me. Later on, when I learned to sing and enjoy the song, "Out of the Ivory Palaces," I comforted myself with the thought that the palaces only appeared to be ivory, and were really made of some heavenly substance. (It was a joy also to discover that this verse in Psalm 45 is translated quite differently in other versions. I am sure now that no elephants will be killed to provide Jesus with those celestial dwellings—in fact, I tend to picture them made of translucent alabaster.)

I could hardly wait to be off for the mosque. But first came lunch and various other unpleasantries at the hotel. I must have been moved by love indeed, for I was by now in weakness and feverish unreality. My back pained a lot, but not enough to keep me grounded. And walking was a real exercise. Yet I had no desire to spend our last afternoon in Cairo in bed. So, at length, we set out for our last tour. Enroute to the citadel we had another fine view of the Cairo tower, which shown with beauty in the afternoon sun. As I admired it, I thought of it as "an ivory tower," a term used for the efforts we make to shut ourselves away from reality and escape into a dream tower of our own making. Someway, all of Egypt's pretentious buildings seemed to me to be such an effort to escape the drab reality of the needy land and her impoverished people. In the ancient past she built fabulous tombs and monuments to the dead. In the more immediate past she built fabulous mosques to the praise of Allah. And now, as the modern age holds sway, Egypt is erecting magnificent buildings in Cairo—showplaces designed to impress the world with her glory. The tower is one such edifice—and is particularly admired by tourists. But I was most interested in that alabaster mosque.

"The Citadel" was built by the famous Saladin. It has a commanding view of the city, and is likewise visible from every quarter. This is because it is somewhat elevated and has very high minarets. I enjoyed the view we had of it from our hotel balcony, and it was indeed interesting to enter its ancient mosque and, later, walk about its walls. We could see the city spread out toward the Mokattan hills. An ancient wall runs all the way from the Nile to the hills. I believe this strategic point was sometimes used for military purposes. But now it is a vast compound, landscaped, and park-like on the exterior, and much enjoyed by the tourists. The interior of the mosque was rather dingy, but vast and majestic. We were obliged to put on the regulation mosque slippers before entering. And there was so much talk and confusion and picture taking that I wasn't able to sense the true atmosphere of the Mosque. Of course once we had passed the outer courtyard, and entered into the mosque, we were quiet, and were taken into a corner to be lectured by the guide. I was in so much misery that I spent all my energy just trying to survive and not show my feelings. So there is little I can report about this citadel mosque.

It was a relief to learn that the Alabaster Mosque, which I am sure has another name too—but I cannot recall it—was nearby. And when we were led into it, I was truly moved upon by the Holy Spirit. I again felt identified with the Moslem peoples of the earth. And I marveled at the impressive beauty of this temple of worship. The alabaster used was mostly a rich golden color, intermingled with white. *Here was a house of worship made all of gold—not of the gold that glitters, but of gold that glows,* and seemed warm and alive! There were many ornate hanging lamps, and we were told that on Fridays—their holy day—and upon feast days, these lamps are all lighted. Once oil was used, but of late mosques are being wired for electricity. And this surely seems a shame to me though the thought of cleaning, filling and lightning a hundred or more hanging lamps is staggering! I withdrew from the guide and the crowd and blessed and praised the Lord, interceding again for all His Islamic people. And I also sought grace to give a clear testimony to all the Egyptians I might contact, before leaving Egypt. (He helped me to do this, I am thankful to say.) I cannot put into words the communion I enjoyed with the Lord in this most enchanting alabaster mosque. (Suffice to say that it lifted me above all my pains and weaknesses.) Neither can I describe the way it appeared to my eyes. It all seems now like a sort of golden dream. I could not help but feel that perhaps in heaven I shall see edifices somewhat like this simple, elegant golden mosque. There were no pictures, statues or other articles to clutter it. *It seemed a fitting place to meet the Ancient of Days*, the God who is enthroned in space. Yet I couldn't help but feel that if I were to enter it apart from the anointing, I would find it overwhelmingly large, empty, impersonal and even cold to my lonely little human heart.

It was a relief to get back to the bus and be able to sit down for a time—for once the anointing lifted, I was utterly miserable. I hoped we were going directly to the hotel. But instead we were taken for a hasty run through the section of Cairo where the bazaar—or *"moskie"*—is located. It was colorful, noisy, crowded and dirty. But some of our number were all for stopping and shopping. Our guides had a little "treat" in store for us—a special excursion into "The Garden of Allah." So *this* was what the guide had meant when he had spoken that magic phrase just yesterday! It surely bore no resemblance to The Wondrous Gardens of Allah into which the Spirit had taken me! But soon I found myself climbing some steep dark stairs and being herded into a hot crowded little room. About 15 of us were thus corralled. The rest, including Dwight, escaped enroute and were presumably having their own exciting time of it in the little shops surrounding this one. We were seated in a circle on chairs, and given a demonstration of the "rare and wonderful perfume essence of Egypt." Several of our number were men, and they could not have been more annoyed. If they bought perfume at all, it would be in Paris, they announced. But I was interested. Perfume essence is distantly related to the rare ointments once sealed in alabaster flasks. And even though there was no myrrh, spikenard or aloes available, there was Attar of Roses, Amber Antique, Lotus and others, all put up in oil—so they could not evaporate. It was evident that until some of us bought, none of us were going to get away—for the door was shut, since another crowd had entered the other room.

We all felt trapped, and even after a few bought, there was no move to release us. Our guide had stepped out, and we had to wait for his return, we were told. It was all very trying and disagreeable, and again tempers flared. The room was closed and hot, and I felt wildly thirsty. I had difficulty in not becoming cross or bursting into tears. Even after we made our escape, we had to wait while others shopped or wandered around. It was very late in the afternoon before I finally collapsed on my bed. There was still no mail for us, and I knew a number of you were planning to write me in Egypt. This gave me an odd disjointed feeling, as though I were in exile. My cold was growing worse, and I was a mass of aches and pains. I felt as though I were in a strange purgatory which would go on and on. I just could not believe that we would soon take flight to Jerusalem and Pentecost.

I may have slept a little before I realized that there was washing and cleaning up to be done before dinner. It seemed impossible to find even an hour for rest! At the table it took all the grace I could muster to be cheerful. But the Spirit came to my aid, and after the meal He opened the way for me to witness to various ones in the hotel. No tour was planned for the evening, for we were scheduled to leave Cairo 7:00 the next morning. This meant that we would have to eat breakfast at the grizzly hour of 5:30! Dwight and I decided to do a little packing before going out—since it would be difficult to manage it all at such an hour. Thus we set out rather late for our last little excursion into the city. We had so greatly enjoyed our walk on the preceding evening, we began another, this time toward the hotel district. But on this night we had no patrolled street. Instead we were constantly jostled and approached by peddlers and beggars. I had all I could do to hold on to my purse with one hand and Dwight with the other. It was not a pleasant experience, and we soon tired of it. So we went into a nice hotel and walked through their shops, buying a few little items as souvenirs. I had one little dream in the back of my mind, and soon I voiced it—I would enjoy going to the top of the tower to watch the view of the city as the tower revolved. Dwight was amazed—for I am not one who usually enjoys heights. But I am sure I felt the leading of the Lord. So Dwight called a taxi and we soon were on our way. It was time to climb Egypt's Ivory Tower and dream about its future!

As the taxi stopped and waited for traffic to clear so we could turn onto the beautiful bridge leading to the island on which the tower stands, a little boy rushed out and dropped a necklace of flowers through the window. The driver rebuked him, but Dwight said, "What is it he gave you?" I picked up the fragile flowers and smelled them, and knew at once that they were some type of jasmine. Needless to say, I was delighted! We found that for this lovely miniature lei, the boy wanted all of $.10! Few things in life have given me such pleasure and sense of love and blessing as that unexpected floral adornment. I felt that our Lord had sent it to me directly from His hand. His Bridegroom love flooded my soul. Meanwhile the fragrance from the flowers ravished my senses and I felt again like a queen, or at least a princess.

The "Ivory Tower" is not of ivory at all, even though at a distance it has that appearance. Its actual covering is of mosaic tile, in blue, yellow, red and white. And truly, illuminated by clever lighting, it is a fairytale spectacle by night, out-"Disneying" our own famous tourist-tower—The Space Needle at Seattle. (Since the two vary in shape, it was difficult for me, when seeing the latter last summer, to judge which is the highest. I was told that Cairo's tower is a few feet higher in the actual tower, but that the point of the needle rises higher than the torch in Cairo does.) In any case, they are both most interesting and dramatic in appearance. And, of course, Cairo had hers first and feels that we copied it. We walked about in the garden beneath the tower for a few moments before getting into the elevator that takes one to the very top in a matter of seconds. Unlike the one in Seattle, this elevator is inside the tower, and I experienced no unpleasantness in the sudden rise. (Also the cost is most pleasant—only $.25 a piece.) Almost before we knew it we were at the top and walking about the tower. The restaurant in which one sits and revolves while eating was closed at this hour, so we made the circle on foot, and marveled at the magical view below.

Cairo is a spectacle indeed by night, and from this height I felt more than ever that it was a land of Aladdin and his many lamps. In the distance we could see the pyramids faintly illuminated. Actually these are very beautiful by night. Not only are they illuminated but the new system of "*son et lumiere*" illustrates and tells the story of ancient Egypt in various languages at stated hours. It was most moving to see them thus from the tower—looking back again on Egypt's past. Just beyond the brightly lighted city, there was a sudden blackness—where the desert begins. To see a great city so brightly lighted, but surrounded by complete darkness, is an unforgettable sight! Then I looked away from all that was below to the multitudes of stars above—wondering just what future is written for Egypt in God's eternal plan. Looking down again, I could see a soft veil over the Nile, and it appeared most bridal to me this night. The wind was stirring, and my jasmine "wedding" necklace was almost overcoming me with its heavy scent. I felt on the verge of ecstasy—but I cannot describe or explain my transport.

This was another interval of the trip which I shall long cherish. For a few minutes we seemed to be projected into a future world of beauty and wonder. Then we turned back into the land of today, and went into the little Coffee Shop where a few people were eating. The atmosphere was gay, and from our table we still had a lovely view of Cairo. But the magic spell was broken. We drank tea and ate a little cake and enjoyed a few minutes of carefree relaxation before we descended the tower. On the elevator there were three very lively companions, who I thought were speaking Italian. So I decided to give the language a try. I said, "*Buona sera,*" to them. And they instantly responded. Then they continued to converse in broken English. When I told them we would be visiting Rome before long, they all united in telling me how wonderful it was, and how welcome we would be. I suddenly felt a premonition that Rome was going to be something *special*—and joy flooded my heart. As we parted, our "*arrivederci's*" rang out musically on that heavy Egyptian air. A taxi was waiting, and we soon were being whisked back to our hotel. I was feeling so happy that I forgot to hold my breath each time we had a near collision. I still felt as though I was in a dream, and that nothing could happen to wake me up. I marvel even yet at how gracious our Lord was in making our last night in Egypt so memorable!

But dawn was quite another matter. We were in the throes of packing before the sun had risen. And at 5:30 exactly we were at the breakfast table. The young waiters were obviously sleepy, as we all were. And there was much talk about how we were being routed out early, only to spend hours at the airport waiting. Those who were to come with the second section were supposedly still sleeping. Once again our group was all mixed up and there was no obvious reason for it. Along about seven we arrived at the Airport, which was already crowded. We hoped it wouldn't take as long to get out of Egypt as it had to get in. And we still didn't have our passports. The fussing was loud and long as we slowly passed down the lines. At just the right moment our travel agent appeared with the needed documents, and we were all hustled through the gates. However, sad to say, our plane was not ready. So we took up our wait. Since all the seating space was in use, there was nothing to do but stand. I was still feeling weak and feverish, but what thankfulness there was in my heart that in spite of the rash, I had not developed measles; and in spite of the fall off the cart, I still could walk; and in spite of my bad cold, I still could navigate! Anything, oh, anything, just to get out of Egypt and on to Jordan! The thoughts of seeing that dear land again made me almost ecstatic.

Suddenly I became aware that a very handsome Egyptian soldier was smiling at me in a most charming manner. (In some ways it is nice to grow older and be beyond the age where one has to be careful of flirtations. I looked every year of my age and then some, and my hair was still razzled-tazzled. Besides, I had a husband in full evidence.) So I gave him a sunny smile and said, "*Sabah al Keyr!*" At this he got up at once and offered me his seat, a cigarette, and a cup of Turkish coffee. This reminded me of the last friendly gestures over a cup of Turkish coffee in Jordan, and tears came into my eyes. Again I felt the Lord showing me kindness through an Arab. Of course, this one too spoke good English. So we had a lively conversation. I felt a real love flowing through me toward him and Egypt and all its peoples. And I witnessed to him also. Since we might at any moment have to go to the plane, I declined the coffee. But his kindness left a warm and a sweet taste in my heart that more than compensated for the lack of that syrupy drink. And the last memory I have of Egypt is seeing him bowing and waving to me as we were on our way to the plane.

For a wonder it was in a hurry to be off! Of course it looked pretty ramshackle, but we were eager to go too. Just ahead of it a colorful little jet had flown in from Ethiopia, and there, painted on it in glorious fashion was the Lion of the Tribe of Judah! Some had assumed that we would fly in it, but this of course was only another rumor. It was a joy, though, to see this symbol of our Lord, and to have a passing touch with another land for which I have a special concern.

We had a smooth takeoff and a surprisingly clear view of Cairo as we circled. My heart was singing and soaring higher than the plane! My head hurt, my ears hurt, my back hurt, and I was getting very hoarse. But the Spirit was not even aware of all this, apparently, and He was giving me wings! I felt an urge to turn to the Word, and as I did, the familiar phrase revived: "He hasted, if it were possible for him, to be at Jerusalem for the day of Pentecost." That Day was at hand! And we were now on our way. Tears overflowed as I realized in spite of all the tests and dangers, obstacles and necessities, we were right on course for the appointed meeting—the first great Christian Assembly in Jerusalem at Pentecost, since the Spirit was originally outpoured. I opened my little Amplified N.T. then, which I always had at hand, and these timely words greeted me: "So, being fitted out and sent on their way by the Church, they went through both Phoenicia and Samaria telling of the conversion of the Gentiles, and they caused great rejoicing among all the brethren." The Spirit changed the locale to Egypt and Jordan, and I felt much joy in the coming conversion of the Egyptians and Jordanians—by faith. I knew that you had truly "fitted me out and sent me on the way," both naturally and spiritually. Then I read on: "When they arrived in Jerusalem, they were heartily welcomed by the church and the apostles and the elders . . ." The Spirit made it clear that the Lord was going before me and preparing the way for all that was to take place there. I knew that I would meet Him, and apostles and elders and precious ones from many places.

I lifted my heart and praise to God, and just at this moment I heard the stewardess say, "Are you Frances Metcalfe?" When I said I was, a letter fell into my lap, and I had the feeling that an angel had delivered it from heaven. It was from Isabelle—and was a beautiful and anointed letter about Egypt, giving Scriptures and confirmations of things the Lord had been impressing upon my heart about how He was going to visit this land and take out a glorious company of people for His name. I simply marveled as I read. No one else to my knowledge received a letter on the plane. Nor did I ever receive another on any plane. All the rest of the letters addressed to Egypt caught up with me later in the trip—I suppose our travel agent had them and carried them along. How this one particular letter, so full of prophetic utterance, reached me at just the moment when the Lord had spoken—and I was leaving Egypt behind facing Jerusalem, causes me still to wonder. I felt He was assuring me that He had heard my prayers—and those of all of us—for Egypt, and that in due time He would answer.

As we neared the Suez Canal, the Red Sea, and Sinai, the view was superb! All along the way we enjoyed seeing what had been veiled from us by clouds a few days earlier. The Gulf of Aqaba was brilliant blue-green, and most wonderful of all was the thrilling view of the Dead Sea, the mountains of Moab, Jericho, and finally JERUSALEM itself, as we circled around like a great bird and made a beautiful landing on its tiny airstrip. This flight I thoroughly enjoyed! And how I praised God that He permitted us to have such a wonderful view. The Word about how He bare Israel on eagle's wings came to mind, and I compared our smooth swift flight with that forty-year-long pilgrimage through the wilderness. In three hours we had covered more ground than they did! In some way this seemed to be the message of Pentecost to me, as we descended into the Holy City. By the power of the Spirit more can be accomplished in a few hours than in long years lived in the energy of the flesh, even though these are directed toward God. How wonderful, how marvelous are the ways of the Spirit! All praise to the God who gives us His wings!

ZION, THE MOUNTAIN OF GOD'S HEART

I t has been said, with poetic insight, that the Holy Land is the Sacred Heart of the World. And the heart of the Holy Land is, of course, Jerusalem, beloved Jerusalem! Jerusalem too has a heart, it is broken and divided—as is fitting for the city where Jesus' tender heart was broken by grief and pierced by the spear. What a sign this is to all the earth—The Holy City torn, divided, double-hearted! How clearly it represents the people of Israel, who, having caused heartache to the Lord throughout the centuries, finally broke the body and heart of His Son on the Cross.

The heart of Old Jerusalem is the Dome of the Rock Mosque, and the area surrounding it, upon which the Temple of Solomon once stood. How proud the Jordanians are that it—and all the sacred places in the Old City—are under their care and authority! They know very well that for the Jews too, this area is the heart of Jerusalem. The Wailing Wall, with the few remaining stones of the Temple, was for generations their most sacred shrine. And the rebuilding of the Temple—where the Mosque now stands—is their most cherished dream.

But the Lord desires to give them a *new* center and a *new* shrine, symbolic of the new heart proffered in the New Covenant, and the new life and destiny to be found in Christ. Hence YHVH has chosen not Mt. Moriah, on which the Temple stands, but Mount Zion, the historical Mount of David, to be the heart of the New Jerusalem. Oh, if only their eyes could be opened to see this truth! "For the Lord has chosen Zion; for He hath desired it for His habitation. This is My rest for ever: here will I dwell; for I have desired it." (Ps. 132:13,14) "This is the hill which God desires to dwell in; yea, the Lord will dwell in it forever." (Ps. 68:16)

Through David, in which Our Lord was typified as King, He made known this desire of His heart. It was David who subdued Jerusalem and built up Mount Zion. It was on this Mount that the Ark first rested, when it was returned, until a Sanctuary could be prepared for it. It was on this Mountain that the glorious Kingdom of Christ was foreshadowed. Because God chose to exalt this mountain above all the mountains of the world, and to make it a type of His eternal Kingdom and place of abode, it has been enshrined in the hearts of Christians in all generations. But especially is it precious and significant to us who live in the light of the Latter Day Manifestations and Presence of the Lord. Our songs and prophecies abound with references to it. We know, of course, that these refer to the New Jerusalem and to the New Mt. Zion, which is eternal in the heavens. Yet the earthly Mt. Zion is a symbol of all this, so it becomes, for us too, the heart of Jerusalem—the place of God's dear desire.

Christians cherish Zion not only for its prophetic, but also for its historic significance. For it was on Mt. Zion that Our Lord ate the Last Supper in the Upper Room. It was here that He gave us the New Covenant, and established His Church. Fifty days later, the Holy Ghost was outpoured on this mountain, likely in the same Upper Room, said to be a comfortable home of John Mark's mother. Hence, for the Christian Church, Mt. Zion is the place of her birth, of her wedding with the Holy Spirit, and of her glory. It is certain that Christians worshiped there off and on during all the early days of the Church, except for times when persecution drove them elsewhere. Therefore, we rightfully claim Mt. Zion as "our" mountain, and revere it in a special way.

The Jews also love Mt. Zion and cherish it. But not for the reasons I have mentioned above. They are not ready as yet to receive Christ and the new heart and Covenant. Their affection for Mt. Zion is because of David, who is greatly loved and honored in Israel. His tomb is located on this holy Mount, and it is the most sacred shrine in Israel. Nearby they have established a memorial also for the millions of Jews who were slain in Germany. Pilgrimages to David's tomb and to this Memorial are constantly being made. The Jews climb the Mount to go to the tomb! The Christian climbs to visit the Upper Room! But both ascend the *same* mountain, the only portion of Old Jerusalem which is in Israeli hands!

You can readily imagine how excited I was when the day finally arrived for me to make my own pilgrimage up Mt. Zion. This expedition was uppermost in my mind, as our plane circled around Old Jerusalem, and I caught a glimpse of the shining tower on the top of Mt. Zion over in Israel. I assuredly share with the Jews their love for David and reverence for his tomb. And I richly share with all Christians their regard for the Upper Room. I realize keenly what Zion symbolizes, and know by the witness of the Spirit that it indeed has been chosen by God, and is heart-of-His-Holy-Heart. But I had, in addition, my own very personal reasons for longing to ascend this Mount. So strong was this longing that could I have visited only one place in the Holy Land, it most certainly would have been Mt. Zion.

Some of you know why I felt this way, for you recall that in 1949 I had a most unusual experience, during the forty days when we were in prayer both afternoon and evening. One night, all unexpectedly, I was suddenly rapt and taken to Jerusalem. This was so vivid a dealing that I can recall it in detail, even today. I was walking in the Old City—and the sights, sounds, even the smells, were palpable to me in an unusual way. I was wearing sandals, and I could feel the old cobblestone pavement distinctly, each step I took. I was clad in a long loose garment, somewhat like we picture the old prophets wearing. And I was strongly wrought upon emotionally, just as you would imagine yourselves to be when at last, you actually set foot in the Holy City where Our Lord ministered and was crucified. (In fact, I was much more wrought than when I actually did walk there in the flesh two years ago.)

As I walked, I lifted up my voice and prophesied. Then, suddenly, I remember hearing myself say, "I *must* go up Mt. Zion, I *must* go to the tomb of my father, David!" This was repeated over and over, and I soon began to climb toward that hill. The ascent from the side of the city is gradual, and only at the end did it become a little steep. I found myself in a very ancient building, part of which was open, as though war or violence had damaged it. Looking down, I was amazed to see candles burning, red damask hangings, and what appeared to be like a large marble or limestone slab, resembling a tomb. Several were kneeling reverently around the sepulcher. My heart overflowed and tears streamed down my face, as I cried out, "O my father, David! O my father, David!" I lingered for a time. Then, without knowing how I got there, I was in another room in the same building. A great awe came over me, and I realized that I was in the ancient Upper Room. There were no furnishings in the room, and it appeared damaged or in the process of decay. But, oh what a marvel to be there, and to think of the great events that took place there! The past seemed to blend with the present . . . and I participated in each.

After I returned from this state of transport, and found myself in the little prayer room of the House on the Mount, in company with several sisters, I was strongly inclined to question my experience. For one thing, I had no idea that anyone even knew where the tomb of David was. And I doubted that it could possibly be located in a building on Mount Zion, in which the Upper Room was also to be found. The more I thought about it, the more incredible this seemed! Yet the visit to the city was so real and moving, that I did not want to doubt the Lord in any way. So I alternated between wonder and torment—not wanting to be deceived, nor the Blessed Spirit to be grieved. For several days I continued to ponder and pray about my little journey. I had been shown prior to this time that someday I would visit the Holy Land in person. And yet this going in the Spirit seemed even more real and vivid than a trip could possibly be (and it was!) I did not see how the Lord could confirm what had happened, since it could not be traced in the Word, and no one I knew had any knowledge about these places. Then, oh wonder of wonders, the Spirit led me to turn to Futterer's pictorial book *Palestine Speaks*. I had bought it years before, but had never found the time to really study it, though I always intended to do so. In just a few moments I found not only the information I wanted, but an actual picture of the building on Mt. Zion where both David's Tomb and the Upper Room are located! It was indeed the building I had visited!

I can't tell you how relieved and overjoyed I felt. To think that the Upper Room was still there—though to be sure it is not the actual house in which Jesus ate, but one built on the same spot. The present building is called the Cenacle, and is all that remains of a Franciscan Basilica, which dates back to the time of the Crusaders. How the tomb of David came to be in this church is a long story—but a true one! The Holy Spirit knew all this, but I could never have even dreamed it. All praise to Him! How I thanked Him and wept for joy, realizing how privileged I was to be borne on His own wings to the very place where He first descended upon the Church. It still fills me with awe! I realized, as I pondered, that this would be the most important place in the entire Holy Land to me, and that I must go there someday without fail.

Now the day to ascend Mt. Zion had come! To be sure, I could not possibly follow the route I had taken in my dream. I would have to go around the other way, by the Valley of Hinnom, which is on the Israeli side. This, of course, was a little disappointing—but getting there was the important thing. And I was on my way! Since we had left Cairo shortly after eight in the morning, we arrived in Jerusalem a little after noon. (The flight took three hours on the D.C. 6. But we crossed a time zone, so that adds up to four.) Everyone was excited and happy to get off that old plane and be "home" again in Jordan. There is something about it that you just can't help loving! We got through customs easily, and were then herded into a waiting room, where there were only seats for about half of us. So I decided to visit my "family," and went all through the airport, greeting and smiling at the friendly Jordanians.

Dwight meanwhile was guarding our luggage, I had heard a rumor that our bus would be delayed, having been dispatched on some other important errand; though what could be more important than transporting us to the Mandelbaum Gate, I couldn't quite imagine! I was glad for this little interval to walk about and say a lingering farewell to Jordan. It was tantalizing to realize that Old Jerusalem was so near, and yet I could not return to it. Now the time had come to go on to the New Jerusalem, and I was both happy and sad! My sorrow was increased when I saw a large sign written in both English and Arabic by the Post Office window. It read, "Positively no postcards with Israeli pictures permitted to be mailed in Jordan." I knew, of course, that no mail passes between the two countries, and no phone calls, no telegrams, no nothing, except bullets. But it hurt me deeply to feel again the hate they have for Israel. And I could not keep from crying a little about it.

I decided to pay a last visit to the Rose Garden where I walked and worshiped four days earlier. There I found a rose very much like a Talisman, and stopped to admire it. The gardener, working nearby, noticed this, and evidently was proud of his flowers, for he came toward me with a wide Arab grin. I greeted him in Arabic, and he was really pleased. He picked the lovely rose and was just about to give it to me when a nice looking, well-dressed man, who evidently had been watching us from a window, approached us and greeted me in English. I hesitated, wanting the rose very much, but not wanting to cause the gardener any trouble. The man nodded his approval, and I thanked the gardener in Arabic. This interested the other man, so we fell into a lively conversation, mostly in English. I asked him how to say rose in Arabic, and he told me, saying that his name was the same. It sounded like *Hanoon*—if I remember right. I said it over and over. Then he introduced himself proudly as the manager of the airport. So, of course, I told him how much we had enjoyed visiting Jerusalem, and being at the airport—which is attractive, though very small. He shrugged and informed me that of course their large airport is now "on the other side" at Lod. I couldn't help but wonder how Jordan could claim this as theirs! But I was polite.

To my amazement then, this dear man invited me into his office to have Turkish coffee with him, after first inquiring if Madamé was traveling with her husband, and including him in the invitation. I was really touched and thrilled! Twice in the same day an unknown Arab had offered me a chair and Turkish coffee. This time I meant to accept, hoping that the dilatory bus would be delayed long enough to permit this pleasure. I rushed to find Dwight, and hurried him along, after asking someone to guard our things for a few minutes. Dwight was really surprised at this proffer of hospitality, but as pleased as I. Of course, he doesn't drink Turkish or any other kind of coffee, but our host was most gracious and got him a cold drink.

The Mandelbaum Gate is only a short distance from the airport, which was very fortunate, for everyone's tensions seemed to mount higher and higher as we approached this world famous portal that is strictly one-way for the tourist, and leads to a land which theoretically does not exist, at least as far as Jordan is concerned. I could not help but feel that it is an apt symbol of the gateway of death, that leads also to a land which many on earth dislike to talk about, preferring to pretend that it does not exist either. There is no other possible way to enter Israel from any of the Arab lands, except through this imaginary gate. Actually it is not a gate at all, and its name was that of a merchant who once dwelt in one of the shell-shattered buildings which line No Man's Land at this point. (I have recently read that this passage has been renamed—"The Way of Simon the Just." If this is true, someone has a strange sense of humor—or lacks one altogether!)

Our bus came to an abrupt halt in front of a small police post, in front of which a frontier pole was stretched across the road. Some 200 feet away there was another such pole and an Israeli police post. We all stopped talking when the bus driver switched off his motor. And one by one we climbed off the bus and entered the station, while two police stood guard. I think everyone felt somewhat like I did—as though we were refugees, doomed to some sad fate. The Jordanians do not refer to Israel by name—the expression they prefer is "crossing over to the other side." Since sometimes we speak of death in the same way, it made us feel funereal when they began to instruct us about how to cross over. We were told that we must go a few at a time. Each one must carry all their luggage in one trip. (This was supposed to include the heavy as well as the hand luggage. Few of us women felt equal to this, since we were very hot and hungry, in addition to being exhausted from our early rising hour and trip. I looked to the Lord for the older women who were traveling all alone.) We were to walk directly to the other side without looking around, talking or creating any disturbance. It would be good to bear in mind that on both sides soldiers would have their guns trained on us during this crossing. Our safety depended on our following instructions exactly. From their tone of voice and facial expressions, it was seemingly evident that the Arabs expected most of us to perish enroute or soon afterward. It would have been hilarious, had it not been so serious! They seemed to feel that to go to nonexistent Israel is a fate worse than death. Or was their sorrow caused by the prospect of a lot of American dollars being spent on the wrong side of the gate!

Our passports were again examined and our visas stamped for our exit. Then, a few at a time, we were escorted out to the pole, and bade a solemn goodbye. Our efforts to be gracious and express regret about departing seemed hypocritical—we were all obviously straining at the leash to get into Israel! So off we started! I think we were already halfway across No Man's Land before I realized that we did not have any of our heavy luggage. I was sick! There was no way to turn back without risking being shot! So on we went . . . without a word . . . without a look or a moment's hesitation. Oh what a relief it was when a genial Israeli officer stepped out and lifted the pole, smiling, admitting us to the Promised Land! He was as welcome a sight as the Admitting Angel of Heaven must be! And there was also another welcome sight! Piled on the ground all around was our luggage! Contrary to all rules and instructions, we had NOT carried our luggage with us. Some Arab angel had actually driven the bus with our luggage over to the Israeli lines and an Israeli angel or two had unloaded it. And all this had taken place without the firing of a shot! It was like a little miracle to us who had prayed about that heavy luggage! But I still can't see how it was done. No vehicles, except official U.N. authorized cars and trucks, can cross between. But one had!

We passed through the Custom and Passport lines again, but it was little more than a ceremony. The officers were most gracious, and they seemed so much like my brothers that I wanted to kiss them. I "shalomed" them all warmly, and they "shalomed" right back. When we were finally all safely over and officially admitted, we were gathered together in a group and warmly greeted in crisp British English by an Israeli officer. He graciously explained about the language, currency, customs, etc. of Israel. And he made us feel that they were simply delighted that all we Christians had invaded their kosher country. In no other country we entered were we treated so royally! And this too made this home coming to the New Jerusalem seem to typify heaven and the sweetness of crossing over at the end of our journey.

There was only one discordant note! One of our Tour sisters, whom I had seen only in passing before this time, was carrying on in a most unladylike, unpentecostal fashion. She had crossed over ahead of us, and her luggage had not been there to await her. No one knew what had happened. The Israeli officer could not possibly communicate with the Jordanian officials—so she was stuck. And she was mad as mad! Although she did not actually swear, her tone and manner suggested that she could do a very thorough job of it, if she once started. I was indeed pained and embarrassed, as were a number of us, that such an unChristlike display took place in front of these obviously well-bred Jews. Yet my heart went out to her too, for she was in a real dilemma!

The rest of us were soon comfortably seated on a bus vastly superior to the type used in Jordan. And while our luggage was being loaded on top, we were a merry little company indeed, in spite of being still without lunch or refreshment. About this time a jovial voice cried out. "Shalom! Praise the Lord! And welcome to New Jerusalem!" I knew at once that the voice was familiar, and laughed to myself. For the Lord had spoken in the Word that morning, that the elders would come out to meet us when we came into the city. The loving brother made a very nice welcoming speech, assuring us that everything was in readiness for our convention. He passed out a pamphlet with information about the city. And there was his well-remembered name—Leroy Kopp! To think that a brother we once knew in close fellowship—he was even present at our original Esther Feast, and portrayed King Solomon in our midst—should come forth to meet us in Jerusalem! In Old Jerusalem I had felt that David met me; now it was Solomon! Isn't that just like heaven, where loved ones long lost will be among the first to greet us? I hadn't seen or heard from him since we moved to the Mount. But he still seemed familiar. Of course, I thought at once of Eddy, too, we used to sing on his radio program! (His son has a work in Israel, and apparently he spends time there also, even though his center is still in L.A.)

Brother Kopp told us that with *"Shalom"* and *"Tov"* we could all get by very well in Israel. He said this was about all the Hebrew he had mastered so far. It was all very gay and enjoyable, and we were in good spirits when our bus driver and guide finally got on and prepared to take off. The guide told us that his name was Bondi, and that he would be with us during our entire stay in Israel. The bus driver's name, he said, was Slomo! Solomon again! We all clapped and shouted, *"Shalom!"* and *"Tov!"* And this seemed to please them very much indeed!

There was an unmistakable atmosphere of vigor and youth here in the New Jerusalem, of which we were keenly aware. The contrast between Jordan and Israel, Old Jerusalem and New, is indeed great. And it seemed that we had stepped out of the past into the future, out of the antiquated into the modern, it was amazing and most exciting! Yet even New Jerusalem is by no means a modern city in comparison with some of ours, and all around were evidences of their connection with the Bible and past generations. I glimpsed what looked like the Citadel of David, and realized that I was seeing it on the opposite side—having already viewed it in Jordan. How near we were to the Old City! And then, just ahead, a little to the right, I saw it—the Sacred Mount! I knew from pictures I had studied, that this was indeed Mt. Zion, in the "city of our solemnities," the Mountain the Lord so loves and desires, the heart of the New Jerusalem!

And, on its summit, unless my eyes deceived me, there was a tower which lifted the Cross of Jesus Christ high over the Sacred City!

> Glorious things of thee are spoken,
> Zion, mountain (city) of our God!
> He, whose Word cannot be broken,
> Formed thee for His own abode.

THEY SHALL ASK THE
WAY TO MOUNT ZION

(Jeremiah 50:5)

H ad it been possible for me to choose the hour for our arrival into New Jerusalem, I would readily have decided upon sunrise, or sunset. For at these times, according to what I have read, this city is marvelously bathed with heavenly hues, and transformed into a scene of phantasmal beauty. There is something about the clear atmosphere of Jerusalem that enhances light and colors and thus lends enchantment to the scene. Since almost all the buildings in the city are made of limestone or other rocks, lights and colors are reflected in a surprising way and are seemingly intensified. Thus, at dawn the city often appears first rose-hued, then golden. And in the late afternoon it takes on a variety of colors—rose, mauve, purple and varying shades. In the moonlight it appears like a large luminous pearl. Its elevation is around twenty-six hundred feet—high enough to keep it cooler than the wilderness that lies to the south and east, and to lift it above the mists and fogs that settle at times along the seacoasts to the west. Since the city is entirely surrounded by mountains and hills, as far as the eye can see, it appears like a beautiful jewel in the midst of a most ornate old-gold setting—for most of the year, at least. During the spring months the old-gold, of course, turns to emerald green, and there are splashes of vivid colors wildly scattered. The wild flower display in Israel is world famous for its beauty, variety and profusion.

Yes, most certainly, I would have loved to have timed our trip so that I might have arrived in early spring, and at the moment when dawn crimsoned and gilded the beautiful city, the joy of all the earth; or if not at early morn, then in the late afternoon, when shadows and subtle shadings of color transform the Holy City, enhancing it with the unearthly beauty. At such times, it is said, one is certain that he has actually witnessed The New Jerusalem, newly "come down from God out of heaven." With deep desire and anticipation I had looked forward to seeing Jerusalem thus glorified in God's own Technicolor at least once during our visit there. Hence, on this day of our arrival, I had to overcome a temptation to be disappointed that we had to arrive at the unglamorous hour of early afternoon. But it was a great relief just to get there at all! Under the glare of the midday sun all things lie naked and open to the eye. There is no sense of enchantment, no aura of mystery in any city at that hour. Its dirt, its shabby buildings, its commercial aspects, are all too blatant. But this was the hour that the Lord had ordained, and I was determined to be glad and rejoice in it. (I will admit, however, that the many delays and frustrations of "Tour Traveling" make it difficult to believe that the Lord is doing very much of the timing!)

When our bus left the Mandelbaum Gate and started its journey through the streets of New Jerusalem, on that hot, wearisome afternoon, I began to feel a most amazing sense of euphoria and excitement. The contrast between the Old City and the new was so great that it seemed we had landed upon another planet, rather than just the other half of a divided city. The streets were clean and quiet, and the buildings were newer, and appeared beautifully white in the reflected glow of the bright sun. It was to me a city of dazzling splendor. I am well aware that the world has many larger, finer, more beautiful cities. But I was certain then, and still am, that—as far as I am concerned—Jerusalem is the most beautiful, most desirable, most heavenly city on earth—a fit symbol and counterpart of The New Jerusalem above. Yet, no doubt I feel this way because it was this Heavenly City that I saw and sensed, as our bus sped along, even though my eyes were eagerly drinking in each earthly street and its buildings.

I could scarcely contain myself when the high tower of the Y.M.C.A. came into full view, though it was several blocks away. I recognized it at once, and knew that our beloved Apostle, Dr. McKoy, was waiting there for us, since he always stays there when in the city. Just as I had exulted at my first sight of Mount Zion, and the Cross of Christ lifted up on its highest point, so now I exulted that the most elegant, beautiful and expensive building in the city is this Young Men's Christian Association building. It is said to be the center of the life of the young people of the city. What a paradox! Directly or indirectly, Jerusalem must honor Christ! This magnificent structure was erected in 1928 by private American funds, and did not suffer any serious damage in the Independence War; though the King David Hotel, which is located just across the street from it, was bombed several times and had to be restored. It is the outstanding landmark of the city, for its tower looms high and is most inspiring to behold. On top is a figure that represents one of the seraphs that appeared to the prophet Isaiah. (Isaiah 6) On the front wall of the tower, in Hebrew is written, "The Lord our God, the Lord is One." This is the *SHEMA*, the fundamental Jewish confession of faith. On the left is Isaiah's great prophecy: "His name shall be called wonderful, counselor, the mighty God, the everlasting Father, the Prince of Peace."

What a witness! The architecture is elegant Byzantine throughout. And on the floor of the vestibule, the famous sixth century Madaba map of Jerusalem is reproduced. In addition to facilities for the recreational use of young people, the building also contains an archeological collection and library. And frequently Christians and Jews gather there for conferences and Bible Study sessions. It is of interest, I believe, that a man in New Jersey believed he was shown this building in a vision, and was used of God to carry through the tremendous task of having it erected. I was almost enraptured by the sight of that tower, and felt a great urge to go to the top of it and enjoy the fine panoramic view of the city that spreads magnificently around it. And I also felt a real pain in my heart, so intense was my desire to visit the King David Hotel nearby. I had read of it again and again, and I knew that many notable ones from numerous nations had passed its portals. Yet, I believe, it was mainly of King David that I was thinking that day. How I would love to visit, yes, even dwell, in his house! I found tears rushing to my eyes at the thought of it. But of course, I knew that Jerusalem was full of people at this time. Not only was our large convention there, but also the innumerable newsmen and spectators who were attending the Eichmann trial. We had heard that all rooms were taken in all the good hotels, and that possibly we would have to stay as far away as Tel Aviv. The thought of such a possibility was truly a test to me, for of all places I wanted to visit and stay in, Jerusalem was chief. And the closer to Mt. Zion, the better!

But the bus sped on, and left the hotel behind. It was obvious that there was no miracle in this matter, and our resting place was not to be the House of King David! In spite of this passing sorrow, I felt positively giddy with joy! And there was surely no natural explanation for this. We had risen that morning long before dawn. We had passed through all the rigors of making an escape from Egypt and landing in Jordan. We had gone through the long delay there, and then made a second even more hazardous escape through the Mandelbaum Gate. We had waited and waited, but we had not eaten and eaten. No indeed, that skimpy continental breakfast and the tiny cup of Turkish coffee had not done much to fortify me for all these rigors. In short, we were all tired, hot, dirty, hungry, and in various states of discomfort. And, above all, thirsty, thirsty, thirsty! My recent fall in Egypt had left my back in a painful state. Even sitting still hurt. And every movement aggravated it. The constant jostling of the bus was hard to take. So I took a moment to thank the Lord that the Israeli bus was far more comfortable than the ancient jigglers of Jordan had been. And there was also a vast difference in the way the driver handled it. Traffic in Israel was obviously quiet and circumspect. So far we had seen none of the mad rushing, jamming, honking, and narrow escapes we had experienced in Jordan and Egypt. Yet I was indeed eager to get out of that bus, get cleaned up, eat some lunch and, above all, REST a little. The painful trouble in my heel had grown worse, and every step I had taken, so far this day, had been an exercise in keeping outwardly cheerful under stress and distress. Between it and my back! And, oh yes, I must not forget to tell you that my cold had grown worse, my throat was very sore, and I was beginning to croak. It was all most annoying and frustrating to make limping, croaking, collapsing entry into the New Jerusalem!

In spite of all these things, the exultant sense of joy and satisfaction continued to uplift me. I was delivered in—not out of—my troubles! And I tell this to the glory of God; for the wonderful quickening of the Spirit, and His witness, is most precious to all of us. And apart from it, I most surely would have been in the same mood as the majority of our companions. I am sorry to relate that most of them entered the New Jerusalem just like they had entered the Old City. They were not a bit happy about this little joyride through Jerusalem. All they wanted was to get out and into a good hotel. Hence, as we began to leave the main streets and to take a narrow road along a ravine, I sensed their annoyance turning to indignation. A murmuring began: "Where are we going now? Why don't we get to our hotel? Probably we're going to be stuck off somewhere in the sticks, etc." Well, after our hotel experience in Egypt, one would hardly blame them. But at this very time, we were actually traversing the road by the Vale of Hinnom, and we knew it not! The bus went slower and slower, for there were people crowding the narrow road. Finally, there was a bridge, and the driver crossed it at a crawl, as though expecting it to give way under our weight. Then we stopped abruptly! And so did the murmuring! On one side I caught sight of a barbed wire fence and military shelters. I was aghast! That meant we were on the border of Jordan. What had happen? We had been all this time leaving Jordan, and now we were back at its fence! For a moment there was absolute silence in the bus. Everyone appeared to be stunned! The tour guide then arose nervously and faced us all. He held a little book in his hands, and began to read from it—to our amazement! And the words he read haltingly sounded utterly foreign to my ears, I was so unprepared for them. He was actually giving us facts and figures about Mt. Zion, and implying that we were about to get out of the bus and climb that sacred Mount!

From where I sat I could see nothing but the valley and the fence. But evidently Mt. Zion was located somewhere nearby. He never got to tell us just where, for he was very rudely interrupted. About a half-dozen men and women began yelling at him at once. I can still close my eyes and hear their clamor. It went something like this: "We most certainly are not interested in hearing about Mount Zion now. What we want is a hotel. We haven't had our lunch! We're starving! We don't want to climb anything, or walk anywhere, or drive either, we just want to get out and get to our rooms." How very hurt, puzzled and embarrassed was our guide. He looked from one to another and then sat down, as though overcome. He began to talk with the bus driver, in Yiddish, I think. And he sounded terribly agitated. It was all so very strange! Here we were near the foot of the sacred Mount. I had looked forward for many years to the time, when, by God's will, I would actually ascend the memorable Mount. Yet here I was, agreeing in my heart, though I said nothing outwardly, with all the rest. I felt that I was utterly unable to take even one step toward, let alone up, that Holy Hill! I was ashamed of myself, of course, for being so weak. But that withering sun, that dazzling glare, seemed to blind me. And my wonderful sense of euphoria had utterly vanished! I was again on the verge of tears.

Our guide then rose, and he too appeared about to burst into tears. He apologized that he had not understood that we had not had lunch. He had supposed that we would want to begin our tour at once, since we had so little time to see the many wonderful things to be viewed in Jerusalem. He told us—rather reprovingly, I thought—that he had supposed we were most eager to go to Mt. Zion. All tourists and pilgrims, he pointed out, wanted to see Mt. Zion first. Whereupon two or three of our number hastened to assure him that we did love Mt. Zion and want to see it; but not until we had eaten our lunch and gotten settled. They told him that he simply had no idea what we had been through that day. Whereupon he agreed that he would now take us to our hotel. And no one dared ask how far that was.

The bus began to back up slowly, inching its way again across the bridge. It continued to back up all the way to the main road. There were many people on it, people who obviously were more eager than we were to climb Mt. Zion. So he did not risk trying to turn around. On we backed! And I can remember yet how odd I felt. To me this was all most significant. We, like the Israelites of old, are not in truth too eager to ascend the Mount of God. We don't want to go up like pilgrims—tired, hungry, thirsty and weak. We want to choose our time and manner of going. Hence, we enter the city fussing and complaining. And we flatly refuse to go to the Mount—even when we sit at the foot of it—until we are good and ready. It was fitting, I thought, that we backed away from it, as one would back away from a King in former days. This, at least, offered a token of respect. Yet, for all I could see how weak our flesh was, I had to admit that I doubted if I could have made the steep climb in the blazing sun, apart from a miracle. (The temperature was around 92°.) I was a little ashamed of the relief I felt when we finally got back on King George Avenue and headed toward the center of the city.

I settled back and prepared myself for a long ride. Our hotel was probably somewhere on the other side of the widely spread-out metropolis. But again we had a big surprise! In a few blocks the bus suddenly drew up and stopped in front of a very nice looking hotel. Everyone came to attention at once, scarcely daring to hope that this was *it*, and that all of us would be there together. But, sure enough, this WAS it! And Bondi very politely—though still obviously hurt and puzzled at our behavior—escorted us in. I took a quick look and was glad I had learned to read Hebrew a little at least. The name of it was The Kings Hotel. What a consolation! If not the KING DAVID Hotel, I was still going to be in The Kings House! And what an inviting lobby we walked into! It was cool and well-appointed, and the elegant marble floors made me feel as if this were a palace indeed. Things were obviously in good order, and in a matter of minutes we had our room assignment, and were walking down another long marble hallway on oriental rugs. When Dwight opened the door, I went straight through the room—dropping my bags on one of the beds—and out onto the balcony. A balcony! I praised and thanked the Lord heartedly, thrilled at the view before me. There, off to the right, was the tower of the Y.M.C.A. Higher up, the tower of the Dormition Church on Mt. Zion was in full view, its cross reaching high, as if to bless the city. "Mt. Zion, Mt. Zion!" I cried it out again and again. Dwight was with me by this time, though the tiny balcony could scarcely hold us both. Then he pointed out to me that to the left, and farther away, the buildings on top of the Mount of Olives were also visible. I could not believe that this was really true. How could the Mount of Olives still be that near? It had taken us so long, and we had journeyed so far to reach Zion, it seemed incredible that the Old City was just across the way. In between lay the walled city, but from our window on the third floor, we could not see it because of the buildings close by. It took Dwight a few minutes to convince me that we were actually facing toward *both* mountains—the Mount of Olives, lying to the northeast, and Mount Zion, closer, and toward the south.

But we were! And I instantly fell in love with that room, without looking to see what it was like. It proved to be clean, comfortable and quiet, and after our time in Egypt it was really heavenly to be there. I threw myself down on the bed, overcome with physical misery and spiritual joy. Strange how perfectly they blend at times!

In a few minutes, Dwight jolted me out of my collapse, by reminding me that we had to clean up for lunch. Apparently the hotel had not expected us for lunch, but they would do their best. And they surely did! After eating in Egypt, we found eating in Israel a feast indeed. Passing through the lobby, on our way to the dining room, I had heard the first real Hebrew conversation ever to fall on my ears! And oh, how it thrilled me! I seemed actually to eat those words too, and it was a joy that I understood a few of them. Once at our table, I decided to launch forth on my own. So I "shalomed" our waiter, and began to "todah" (thanks) at every opportunity. To my chagrin, he was most unresponsive, and even appeared embarrassed. I suppose my Hebrew must be pretty queer and was about to retreat to English, when another waiter came smiling and bowing to our table. I decided to try again, and this time my Hebrew hit home. He at once began an animated conversation with me in Hebrew, to which I kept saying, "tov! tov!" (good, good.) He finally sensed my limitations, when I managed to stutter out, "Awni medaberet Evreet me-ought me-owed" (I speak Hebrew

341

only a very little.) He was more than eager to instruct me! (I later found out that this was the headwaiter, and due to his attention, I had *"ne plus ultra"* service all the while we remained in the hotel.) He told me that the cook was Hungarian, and that some of the waiters do not understand Hebrew, but only Yiddish and other languages. He also explained that in Israel everyone was learning Hebrew together. And he was most pleased to find out that I had been studying it also.

We had been instructed to gather in the lobby immediately after lunch so that we could begin our afternoon tour at once. It was by now very late, and obviously our tour would be short. A number of our companions said they just couldn't make it, and went to their rooms to rest. I will confess that I was tempted to seek the company of my own nice comfortable bed, after a real bath. Egypt's dust had settled pretty heavily upon me, and was by now most uncomfortable. This was one of the great frustrations of the Tour. To be in a beloved place, to long to visit the memorable sights, and yet to be dirty and exhausted! Well, after coming all those miles, through all those hazards, I was not about to miss a possible visit to Mt. Zion. So I hurriedly prepared to go. Dwight likewise never wanted to miss something.

To my dismay, the bus did not turn again toward Zion, but in quite the opposite direction! And soon Bondi began to point out the various buildings and landmarks of the city. As we circled yet higher, there was a little field, which once was a woods, and from which the cross of Christ was supposedly cut. There is a small monastery close by. Bondi told us this rather reluctantly, and everything else he talked about was decidedly Jewish. I sensed that Bondi did not quite know how to take us Christians. At the end of the Jaffa (Joppa) Road, to the left, a large new convention center was called to our attention. In this *Binyanei Hauma*, (meeting place of the people), our own Pentecostal Convention would be gathering on the morrow. I felt a great sense of joy at the thought, but I realized that the one who wrote that we would be meeting "a stone's throw from the Upper Room" had indulged in oriental hyperbole indeed, or else had not looked at the map! Dear Mt. Zion was nowhere in sight!

We were also shown the Hebrew University campus, which stretches southward in a beautiful mountainous landscape. It has various impressive buildings, including an Archaeological Museum, which houses some of the Dead Sea Scrolls. We would have been most welcome to have visited the University, but of course we had no time. We sped on, passing hill after hill. Jerusalem is surrounded by mountains too numerous to name or number, it seems. On one side we glimpsed Mt. Herzl, and Memorial Mount close by. Then soon we were whirling around curve after curve, as we descended into a charming little vale, beautifully terraced with olive trees and vineyards. This, Bondi told us, was the famous Ein-Karem, the birthplace of John the Baptist. (Some authorities do not agree about this location.) I had read about Ein-Karem (Spring of the Valley) in several books, and was most eager to visit this little "city of Judah" in the hill country, where Mary visited Elizabeth so long ago. The Arabs honored Mary by calling it, *AIN SITTI MIRIAM*, (Fountain of the Lady Mary.) Two churches here claim distinction. The Franciscan Church of St. John is supposedly built over the grotto in which John was born. The present building houses remnants of ancient mosaics and other artifacts. The Church of the Visitation is also Franciscan, and is located near the spring. This too is of historic interest, containing reminders of the Crusaders. I had hoped to visit these places, but I soon realized that our bus was not even going to slow down, let alone stop! One of men had called out, "We don't want to see any Catholic Churches." And there had been a chorus of agreement from our group. So all I could do was pay a fleeting tribute to dear John, Elizabeth, Zacharias and Mary, and whirl on up and over the hilltop. My pain in this rudely abrupt visit was soon supplanted by surprise and fear. Our driver suddenly turned off the road and started up another hillside on what appeared to be a wagon track! (And here I had just begun to feel safe with him, and think how nice Israelis drive!) It appeared a foolhardy thing, and several of the men began to talk about it. Even in Jordan the driver at least had stayed on the roads! We all looked out the windows in consternation, and Bondi broke into a wide grin. He assured us there was nothing to fear, Slomo was the very best driver in Jerusalem, he told us, and could handle the bus safely under all conditions. I began to feel that perhaps this was their naive way of getting our attention and admiration. It seemed to me that they were showing off like little children. We finally crawled over the top of the hill and found another orthodox road.

Nearby was a school where real children were outdoors "showing off." Their dress was most interesting, for the little boys were wearing crowns and carrying tiny scepters, and the little girls wore flowers and ribbons and carried banners.

Some of our men wanted to take a picture about now, so the bus finally stopped. I was delighted to get a closer view of the children. And I asked Bondi the meaning of this celebration. He appeared completely puzzled and said he would ask one of the children who was approaching the road. I felt that this was in honor of King David and Pentecost. And sure enough, this proved to be so. Bondi explained, rather haltingly, that it had something to do with King David—and a holiday. I supplied the details. "Of course," I said, "David was born and died at Pentecost, and the Feast is at hand." At this Bondi was really surprised. "You know about Pentecost?" he said. "Yes indeed! And also about dear King David!" This seemed to touch him, and I felt a real love coming in my heart toward him, and a prayer for his salvation.

As we drove on, I found myself exulting like a child and feeling that David had come forth to meet us in a special way, in that little boy who ran toward the bus. (I can still see his proud little Jewish face and the little crown.) I had no need now to ask Bondi if we were now finally on the way to Mt. Zion, for my heart told me so. After all, no one knows the way better than David, and he had met us! Soon, with our beloved King, I would be ascending the sacred Mount, and by His grace I would stand in the Upper Room!

It is just about three years since our First Missionary Journey. So it has been a joy to live this portion of it over again, as I have prepared this writing. It was a special pleasure to receive a letter from SLOMO, our bus driver! We are invited to his son's Bar Mitzvah and birthday party! Since Slomo means Solomon, I feel that I have heard again from David, through a son! This little earthly sign will likely thrill you too, so I have related it.

THEY WENT UP INTO
AN UPPER ROOM

When the Eve of Pentecost was almost come, I stood at the foot of Mt. Zion, looking up at the banners which were arrayed just above. Every one of these was an Israeli flag, with its proud Star of David. There were a number of pilgrims coming and going, and some of them were dressed in a colorful way. No doubt most of them were Jews, but they were from various nations, and many of them looked much like the Arabs who thronged through the streets of Old Jerusalem. I fell behind our crowd, wanting to pause and praise before starting my long-dreamed-of ascent up the sacred Mount. I tried to realize that it was to this very place that Jesus and the disciples had come, on the night of Passover. And that when the astonished disciples had returned from Mt. Olivet, after watching our Lord ascend into the heavens, "they went up into an upper room . . ." Now, praise God! I too was going to make that climb and stand on the holy Mount! I thought also of you, my darling sisters. I truly longed that every one of you might be there in person. For a brief moment I tried to picture us with OUR banners, going up in procession, singing, praising the Lord all the way! It was a beautiful scene in my mind! But when I opened my eyes I saw only our Tour members streaming out behind the guide, talking and jostling each other. There was no outward sound of praise at all!

I was glad that Dwight had paused to read the sign posted where the path begins. It proclaims to all that this is Mount Zion, a sacred place. Women are expected to be modestly dressed, and men should cover their heads, according to Jewish custom. I was indeed grateful for this attempt to instill a sense of reverence in each pilgrim's heart. But I fear that few of our number even read the sign, let alone thought that it referred to them. (I confess that the lack of reverence one finds in many of our "people" is very trying to me at times.) But Dwight, bless his heart, had gone over to a little booth and purchased a *yarmulke* (which is a little black skullcap worn in a synagogue and for all sacred occasions.) So he and I started up together. But soon he got interested in picture taking, and I was content to become the tail and to lag behind all the others, as I attempted to commune with the Lord.

We kept meeting pilgrims descending, and since the road was narrow, my contemplation was constantly interrupted by a happy, "*Shalom!*" or "Hello!" Everyone seemed to be in a holiday mood. And I found a joy in greeting them warmly. The sight of Dwight's yarmulke identified us as Jews, I suppose. And I felt like one indeed. We came soon to some rather steep stairs, and I had to pause for breath. It was late afternoon, but still quite warm. Since Zion is the southwest hill of Jerusalem, the sun rests on it at the close of day. I could look back and see the Valley of Hinnom on the one hand, and the desolate No Man's Land on the other. Jordan curves around this mount in an odd way, and Israel has only a portion of the land.

On a little higher level we came upon a strange little plot of ground—a Christian cemetery! The dear saints who were buried there died centuries ago, I am sure. And I was strangely moved, thinking that their bodies had hallowed this Mount in a special way. High up, of course, lies the tomb of David, and a memorial to the two million [*sic*] Jews slain in Germany. But here was a little handful of Christians—and in such a prominent place! Directly above them, on another level, stands a huge *menorah*. It is not an elegant candlestick, for it is badly weather beaten. Nevertheless it is impressive, even in daylight. At special times it burns brightly all night long on Mount Zion, and must be seen for quite a distance. The site of this menorah reminded me of course of our own Candlestick, and of how we were led to climb on a rock and lift it high, when we first moved to the Mount; and of how it burned brightly, and in the darkness of the night gave forth a beautiful light. I recalled crying then, "Arise ye! arise ye! get ye up unto Mount Zion, unto the Lord your God!" I wanted to cry it again. I thought too of the large candlestick that was lighted in the Court of the Women during the Feast of Tabernacles, and how it gave light to all Jerusalem. (I learned later that on that very night, and also on the other two nights of the three day Feast of Pentecost, this menorah was kept burning all night, and that the Torah was read both in ceremony and by individuals, until dawn.)

The next thing I knew, I was inside a little shop, where our Tour members had gathered to do their usual souvenir hunting. I was shocked by this, it seemed so out of keeping with the meaning of Mount Zion, so I soon made my escape and wandered a bit on my own. There was a tiny, poorly kept park, off to one side, and lunches were being eaten, children were playing, and everyone was visiting as though at a picnic. Everything on Mount Zion looks rundown, desolate and dejected. Fulton Sheen says, "The Mount is occupied by the remains of the Crusading Franciscan basilica; weed-grown courtyards, beautiful gateways with dog tooth molding, and vaulted corridors, are all permeated with an air of decay and abandonment." And this is the way it impressed me too. This condition increases as one moves on toward the cenacle and the tomb of David. There is heavy barbed wire all along the pathway, and the ruins of a number of buildings and walls remain just as they were at the end of the war. I hurried through this, and waited at a little gate which separates the sheep from the goats, so as to speak. The dear goats, the Jews, turn right at this place, and go on to their Memorial rooms and the tomb of David. The Christians go on through the gate (of the sheepfold?) and follow a little walk to the entrance to the cenacle. Our guide, however, turned aside and rang the bell at the door of the Dormition Monastery, which stands, according to tradition, on the ground where John's house once stood, and where Mary, our blessed mother, "fell asleep." Hence the beautiful name, *DORMITION SANCTAE MARIAE*—Latin for "Sleep of St. Mary." Before any of our companions had time to protest, an aged German Benedictine Monk was bowing to us and welcoming us to his beloved shrine. He looked for all the world like Johnny! And he talked like him too! So I felt right at home. It was obvious that some of our brothers had thought we were getting close to the Upper Room, and they were most reluctant to enter another Catholic Church. But in we were herded. Bondi, of course, remained outside, wearing a typically Jewish expression of superiority mingled with tolerance.

This abbey church is one of the most beautiful sanctuaries built recently in the Holy Land, it is said. And it is just about as old as I am! On this ground the Byzantine church of Our Lady of Mount Zion formally stood. And this, supposedly, was where John the Beloved's house was located, close to the Upper Room in the home of John Mark. I cannot recall the interior of this church plainly, but I was sweetly moved on in the Spirit. I had regretted not being able to go to Mary's tomb in Jordan. But here I was—surprise!—in the place where she died! I was fascinated by the unusually beautiful mosaic floor, and as I stood meditating, a dear colored brother gradually approached me. He was a minister from San Diego, and was one of the nicest brothers in the tour. By this time most of the others had gone downstairs to the crypt below. Only he and I lingered. When I looked into his face, I saw something very beautiful in his eyes. He began to tell me that he wished he understood more about the symbols in these churches. He said he had noticed that I was more reverent than the others, and seemed to have a knowledge of the Holy Land and these shrines which he lacked. Very humbly he asked me if I could explain the floor mosaic to him. Well, this was really a challenge! I had to look to the Spirit in haste. But He was right at hand, and to my joy, He really anointed me to explain the mosaic to the brother—and to myself, at the same time!

I have a picture of the floor before me now, so I can tell you that it is circular, and that the signs of the Zodiac form the outer circle. I explained to him that these are mentioned in the Book of Job, and were known to the ancient Patriarchs, who read the signs in the stars, in which the Gospel is written. I told him about the various tribes using these signs on their standards. And that our Lord comes forth as the sun, in His course, and passes through the entire circle of the heavens, just as the sun passes in turn through each sign of the Zodiac—a universal ministry! I had never thought of that before, myself! The brother seemed thrilled, so I went on. In another circle, inside this circle of the months, were the names of the twelve Apostles and St. Paul. Very ornate designs, including faces of angels, fill in the circle. The names of the four major prophets are in a still smaller circle. In the center is a symbol of the Holy Trinity—three interlocked circles and in each is the word, HOLY, in Greek! Around this is a huge sunburst design. A passage from the Book of Proverbs is written in a circular inscription: "I was set up from everlasting, from the beginning, or ever the earth was. When there were no depths, I was brought forth; when there were no fountains abounding with water, before the mountains were settled, before the hills was I brought forth." This, of course, I could not read, for it was in Latin.

Our companions began to come up from the crypt, so I had to hasten to get to see it at all. It is a solemn and majestic rotunda, which is occupied by a life-sized figure of Mary, lying with folded hands as in death. It was so lifelike that I was stunned for a moment, as though seeing her actual body. I felt as though I was at her funeral! Above her bed is a decorated cupola. In its center Jesus is depicted inviting His mother to heaven. In a circular inscription words familiar and loved by us are written: "Arise, my love, my dove, my beautiful one and come away." Medallions in the circuit show famous women of the Bible: Eve, Miriam, Jael, Judith, Ruth and Esther. (There is another tradition that Mary died in Ephesus in the home of John, when he was residing there. And it is doubtful if either tradition can be proven. I am thankful, in any case, that there is a Christian Church and Cross at the pinnacle of Mt. Zion.) Some of our companions seemed disgusted with this crypt and statue and the church and everything. But there was something about that aged Benedictine father that precluded criticism or levity. So they held their peace.

By the time we were out in the open air again, the sun was beginning to lower, and there were lovely shadows on the buildings. I looked up to the tower, from whence the Cross is lifted high, and wished I could climb up there and have that sunset view of Jerusalem, for which I longed. But this was the moment ordained for our visit to the Upper Room, and it was just a few steps to the stairway. At the top of these ancient stairs there is a roofless anteroom, through which one crosses to the cenacle itself. The meaning of this word is "dining room"—retaining the tradition that Jesus ate the Last Supper here with His disciples. The present building was constructed as a Christian Church in 1130, by the Franciscans, to take the place of one built in the fourth century. In the 16th century it was seized by the Turks and turned into a Moslem Mosque, sad to say. Only in the last few years has it been in the hands of Israel and accessible again as a Christian shrine.

Many scholars say that this location is the most authentic in Jerusalem, next to the location of the Temple. In the days of Jesus, this portion was within the city walls, and was occupied by the well-to-do. The tradition for this site goes even further back than A. D. 400, when Epiphanius wrote that Hadrian, in A. D. 135, found the Christians in possession of "the little church" in which the Apostles assembled after the Ascension of our Lord. It stood in the quarter of Zion that was spared when the city was destroyed. But I needed no statistics or proof from scholars to confirm to me the locality of the Upper Room, for the Spirit Himself took me there many years ago (as is recounted in a previous episode of this trip) and made it plain to me that this was indeed the Upper Room, and that David's tomb was nearby.

Actually, his tomb is in the same building, and beneath the Upper Room. But it seems completely separated, for Christians have to enter by another stairway now. Part of the building was either damaged or closed off so that we cannot go from the cenacle to David's Tomb, as I did in my visit there in the Spirit.

I could not help but pause for praise before entering the Upper Room. So, again, I was bringing up the rear in an inglorious fashion when I passed through the door. Whatever I expected or hoped to feel was definitely NOT to be! Our companions were all talking and walking around, taking pictures of one another and acting quite like this was just another tourist attraction. There was no atmosphere of praise or worship, or even of rejoicing in the Lord. I wanted to turn and run out. But I realized that this was IT—the Upper Room—and I might never enter it again! So I lingered. After some of the picture taking was accomplished, Dwight suddenly spoke up loudly. I can't remember his exact words, but it was something about this being a very sacred place, and didn't we all feel that prayer would be in order? Without waiting for agreement, he called upon one of the leading ministers to pray—and pray he did, though he seemed a little reluctant. At that moment, praise the Lord, the Spirit spoke to me that I MUST return to the Upper Room, and keep a tryst with the Lord, just as I had returned to Gethsemane. So my disappointment was turned into joy and anticipation. And I was willing to follow the others on down the stairs, where Bondi waited to take us to the tomb of David.

It is a mystery to me—as well as to countless others—that the tomb of David is located in part of the same building that houses the Upper Room. Yet this seems to be well-authenticated by tradition. When Peter preached on the Day of Pentecost, he referred to the sepulcher of David as being present and known in their day. Earlier it had been robbed by Herod the Great—when he robbed other tombs also, to provide for his great building program. We know that tombs were usually outside the city walls, but in the days of David the walls were in a somewhat different location. David loved Mt. Zion. His palace and earlier buildings were most likely located on the southeastern portion of the city, where Zion had its beginnings, it is believed. However, as it grew, it gradually spread to the southwest portion and to the high rocky ledge which is now Mt. Zion. It is therefore to be expected that he would be buried there in a place of honor.

When Jerusalem was destroyed in 70 A.D. a portion of Mt. Zion was not harmed. In that portion was included the location of the Upper Room. So it is believed that there was a Christian church on this site from the time of Christ onward although there were interruptions when the believers had to flee from the city. This makes it the oldest site of the Christian Church in the world. It is reported that along about A.D. 1158, a wall of this present building fell, and that in digging to rebuild this, the workmen discovered a cave. They followed this cave and came to a large regal tomb. There was good evidence that it belonged to David. In any case, this is probably the most sacred shrine in Israel, and it is regarded with much deference. The Jews believe that David was born and died at Pentecost, so at this particular season the number of visitors is increased. We had to pass through a crowd of Jews to get to the little room where his large cenotaph lies in state. It is under the Upper Room and somewhat to the east of it. The larger portion of the building is devoted to a memorial for the Jews slain in Europe under the Hitler regime. This is called *Martef Hashoa* (Chamber of the Martyrs.)

I was surprised to find many small candles burning here, in much the same way as the Catholics light vigil lights in their churches. The air was smoky, and depressing. In the ancient room in which the cenotaph of David lies, there was little else except, I believe, a large scroll, the Torah, and some damask hangings. It is made of stone and has silver Torah crowns. The actual cave where David was buried lies under this. In Nehemiah 3:16, we read that there was a pool nearby, and this is doubtless the Lower Gihon which is a little to the west on a lower level. A few candles were burning and some elderly Jews were kneeling reverently. Others merely stood looking around, or whispering, and children strolled in and out. It was all rather odd to me, and so very different from the time the Spirit had taken me there to honor David. I had been filled with awe and wonder then, and had looked down upon the tomb as from above. Yet, praise God, there was no doubt about it, this is what I had seen in that amazing visitation! But the great sense of awe I felt then was lacking. I felt empty, dry and a bit disappointed. It was hard for me to realize that here beautiful David had grown old and feeble, that his glorious voice had grown weak, and then was silenced; and that he had been buried as Jews interred their dead, without embalming, and had doubtless quickly decayed and returned to the earth. It is true that he once had sung, "My flesh also shall rest in hope, for Thou will not leave my soul in hell, neither wilt Thou suffer thine Holy One to see corruption." Yet these things, and according to Peter's sermon on the day of Pentecost, were spoken of the resurrection of Christ. Peter also said, "For David is not ascended into the heavens . . ." But he spoke, as we know of his physical body, not of the soul. We—praise God!—have in vision seen David in the heavens, leading forth the great heavenly choirs, in worship and song. We have had most wonderful communion with him again and again. We have heard his voice and tuned in his songs, and had even "captured" some of them.

As I stood there close to his tomb, I felt this glorious truth, "Because Jesus lives, David lives." Our God is not the "God of the dead, but of the living." And I recalled the wonderful year of the GREAT FEAST, when the Lord spoke clearly about David coming into our midst, and these words were given: "This love lives again, after 3,000 years." This man lives again also! He has indeed become to us a father, a brother, a lover and friend, and we have known that we are part of the blessed "Tabernacle of David," which our Lord promised to rebuild in the latter days. He is our KINGDOM father, just as St. Paul is our CHURCH (*Ecclesia*) father. And we can love and honor and KNOW him in a real way, and not as those who are acquainted only with the historic David.

It was time now to turn away and follow Bondi on out and down the mountain. The day was drawing to a close. I saw none of the beauteous colors I had hoped for. Evidently this was going to be a rather ordinary sunset! The pilgrims were still thronging up, as we worked our way through them and on down to the little park, with its scraggly lawn and cypresses, on past the large menorah, and the little cemetery where saints lie entombed beneath ancient olive trees and cypresses, on down the stairs and into the lower path. I paused and looked back for a last salute to the Sacred Mount. It is a lowly one indeed, as well as a holy one—little among the mighty mountains of the earth. It is lowly enough for children and elderly pilgrims to climb. And it is friendly, holding to its heart its lowly dead, in the little cemetery, as well as its regal dead in the tomb of David and those of his household. The Jews, as I previously wrote, go up to visit a tomb, whereas the Christians go up to visit an Upper Room. If, as the Jews believe, David did die at Pentecost, this was indeed a prophetic sign of his identity not only with Christ, of whom he often sang in prophesy, but also of his identity with the Holy Spirit, of whom he also sang in remarkable inspiration.

It is said that the Jews believe that David is highly exalted in heaven and that he will have a throne second in splendor only to that of YHVH. Surely the day is drawing near when their eyes will be opened to behold the Son of David, and their hearts will finally understand that the splendor and glory of the Kingdom of David was a prophetic sign and a prophecy received of the everlasting Kingdom of his descendant, Jesus Christ! Then they too will be going up Mt. Zion to the ROOM, as well as to the TOMB!

As you know, it is now exactly three years ago that I climbed Mt. Zion and went "up to the Upper Room," as well as down into the tomb of David. After hearing from Slomo, our bus driver, I was doubly surprised to receive at this time a letter from another Slomo, a dear friend of Dr. McKoy, and also of our brother John. This Slomo has been carrying on an apostolic work in Jerusalem under great difficulties. The Lord has spoken to me about him several times, when in prayer, and I have felt that recently He has really "given" him to us to carry in all our hearts, and to assist in financial ways, as He leads. He made it clear to me, when I was in Jerusalem, that we should have a representative in Israel—one who is truly Pentecostal and Apostolic. For a while John filled that place. Now, I am sure that the Lord has chosen Slomo. I feel that I should have written to him before this. But, since I did not, the Spirit moved on him to write to us. He made no request for anything material, but only for our prayers. And I am sure you will all thrill to his testimony, which I will be sharing with you in a separate writing. As his letter has been made available to the sisters coming to the Sanctuary, several of them have told me how the Spirit moved and witnessed concerning him. This is a confirmation. I might add also that he was *filled with the Spirit* one night as he and John were praying ON MOUNT ZION!

THIS IS THE DAY THE LORD HATH MADE

One of the most bewildering aspects of modern touring—wherein a few short hours one is propelled from continent to continent, in and out of several time zones—is a growing sense of disorientation to familiar time and space. It is a most disconcerting feeling to wake up in a strange bed, in a strange room, in a strange country, and to wonder where you are, how you got there, and what time it is. And I had undergone this experience several times during the nine days since leaving home. But when I awakened in my comfortable bed in Jerusalem, on the morning of May 19, 1961, I knew exactly where I was and what I was there for! And I knew what *day* it was—"This is the day that the Lord hath made, let us be glad and rejoice in it!" My heart was filled with praise and thanksgiving that at last the time of Pentecost had come, when the World Pentecostal Convention would convene in Jerusalem, and from all over the world Spirit-baptized Christians would gather in the Holy City, as Jews gathered there on that memorable day when the Spirit was outpoured in the Upper Room. It was a great and significant day indeed. And I was actually there and able to be among this favored company!

I am sure you know that my thoughts turned toward you, as well as toward the Lord. For you dear ones had made my going possible with your prayer, faith and gifts. And in particular I thought of those of you who had prayed and prepared for me to go from the very time you first heard that there was to be such a trip and gathering. Your desire for me to be present, your faith and hope about it, had strengthened and heartened me again and again over a period of many weeks and months of preparation. Yet even with all this assistance, it still seemed like a miracle to me that I was really there. I briefly reviewed in my mind the many obstacles and difficulties involved in my going—including the last agonizing test when the flight was canceled. Then too there were our many escapes from dangers along the way since we had started out. And especially the tests in Egypt. It was something to get out of there alive! (At least it has seemed that way to me.) But here I was—and right on time! I hoped you all knew it and were rejoicing on this day with me.

It was indeed exciting to think of the coming meetings. And there was another special pleasure to anticipate this day—a reunion with dear Dr. McKoy! We had contacted him by phone the preceding evening after dinner. And he was as delighted as a child to know we had arrived. He proffered a very warm invitation to dine with him on Friday night at the King David Hotel! And that was a special token to me—for if I could not stay in the house of David, I could at least sit down at his table. It seemed to me that the Father Himself had welcomed us and issued the invitation—for I often feel God as Father in our "beloved Apostle abroad." As I prayed about this and tried to be calm enough to meditate, I recalled that this day had another special purpose too. It was Sabbath—and Sabbath in Israel is always special. Some weeks earlier, I had read a very interesting article about the Sabbath and a special Synagogue lecture and tour which is conducted every Friday afternoon in Jerusalem. However, I had forgotten about it—in all the excitement of our departure. But now the Lord reminded me that when I read it I had been stirred to pray that I might make this tour and visit the *Mea Sherim* district of the city—where its very heart is to be found. I had felt at that time that the Lord was most eager that I do this, and that once there, it seemed most difficult to try to fit it in. I realized that this would be our only Sabbath in Jerusalem, and that I would have to go this very day, if at all. My problems about keeping this appointment fell like a shadow across my heart. But I purposed to obey the Lord and find a way to fit it into this beautiful, wonderful day. There was one other appointment which I must also fit into the day—decreed by Dwight, if not by the Lord. He had practically issued a decree that I not go another day without buying some glasses. He was really stirred up about it after observing my discomfiture in trying to unpack and get organized the night before. In trying to use his reading glasses, I was frustrated—for they were much too strong for me, and everything I looked at seemed to rise up and almost slap me in the face. But if I took them off I was practically groping. And I kept losing things and getting the wrong things in hand. Of course I was very tired by the time we had finished dinner and could go to our room. And the lights were dim, even in this fine hotel. (This seemed to be the case everywhere we went. I supposed it helps conserve electricity.) He knew too how hard it had been for me to take notes or to write cards and letters. So he felt the time had come to take action. Even now, as I was lying in bed praying and praising, I knew he was out taking a walk trying to locate an oculist. Sure enough! In a few minutes he returned and said he had an appointment all set up, with the aid of the hotel clerk. So it was time for me to rise and begin this never-to-be-forgotten day.

I dressed quite rapidly for me, for I discovered that I was surprisingly hungry. We had eaten sparingly in Egypt; but now we felt at home—and the food was so good that we really wanted to eat. I was not happy about those continental breakfasts however, for a roll and a cup of coffee seem like a skimpy way to start the day. But this is all our tour contract called for. Picture our delight then, when we found breakfast beautifully laid out for us on a large buffet table on the picturesque balcony of The Kings Hotel! It was a typical Israeli breakfast too—which means that there were tomatoes, cucumbers, onions and greens, as well as fruit juice, along with yogurt mixed with sour cream (mmmmmm good!), cottage cheese and other cheeses. Oh yes, and gefilte fish! After we had helped ourselves to what we desired, we sat down and were served hot rolls, eggs and coffee or tea. It was really a most enjoyable meal. And our friendly headwaiter hovered about the table, encouraging me to talk to him in Hebrew about the various foods. I suddenly realized that I was feeling much better than I had for days. I was still a bit hoarse, but the worst of my cold seemed to be over, and my back was not so sore from the fall. I had changed shoes—having use my walking sandals too much, I guess. I felt the Lord showed me that I should change my shoes often—a thing I had known but forgotten about. Anyway, I could walk without so much pain now—in fact I felt almost like running about and leaping for joy. Oh what a beautiful morning! Oh what a beautiful day! I had a wonderful feeling that everything was going God's way!

As soon as breakfast was over we started out for the oculist. Our bus was to arrive shortly to transport us all to the opening meeting of the Convention. And we could not return in time to catch it. So the clerk told us where to get a public bus, and what to tell the bus driver. It was a real thrill to start out together, just the two of us, and walk hand-in-hand through New Jerusalem streets, as we had in the Old City. And it was only a few blocks to the oculist. On the way we passed the office of the Tour Agency which was looking after us. On an impulse I went inside and inquired about the Synagogue Tour. I had tried to prepare Dwight about my intentions, fearing that he would not be willing for me to go anywhere alone, as he had not been in the other places we had visited. But he seemed not only willing but glad for me to go.

He agreed to go to the afternoon session of the Convention and I could take the tour. And the office man assured me that the tour would be held as usual at three o'clock, and said our guide could tell me just how to get to its starting point—or the hotel clerk would know. This set my spirits soaring, and I almost flew into the oculist's office—where I learned that I was not quite as blind as a bat. But I needed glasses all right.

He and his young wife were most friendly, and it was only a matter of minutes before my eyes were examined and all fitted out with reading glasses! It was not at all like the examination one gets here in America, I can assure you he simply held up lens after lens until I told him which seemed best. And that was it! (I still have these glasses, and they are far stronger than I need. But I guess my eyes were pretty tired.) Dwight got me a little chain to wear around my neck and fastened these on—and he instructed me never to take it off during the day. Thus they would be always near. There was opportunity to speak a word of witness for Christ to this dear couple and to converse a little in Hebrew. They assured me that I spoke it FINE! Bless them!

By this time we were both feeling more at home than in Idyllwild—and excited at being out on our own—away from the tour group. We had a lot of fun looking in shop windows, while we waited for the bus. And when it came time I informed the driver to be sure and let us off at the Binyanei Hauma. He merely nodded, and apparently knew no English. We enjoyed jogging through the streets and observing the people. Truly one sees a great variety of races in Jerusalem—it is most interesting! And we smiled at all the people on the bus and acted like we knew them. I suppose this is one reason a little blonde Jewish lady jumped up eagerly when we approached our destination. I was getting a little concerned about where to get off, for I could not understand anything the driver called out. I wondered if he spoke Hebrew either. So when I asked him about it, this lady called out, "I know the way! I show you!" And she was all smiles. When we got off, before I could say anything, she announced, "I Miriam, I Christian!" And she hugged me with an almost fierce bear hug—little bear, that is. (She was less than five feet tall, and quite chunky.) That was really a hug to be remembered!.

Miriam's English was broken, but easily understood. And as soon as I—somewhat breathless from the hug—gasped out, "I am Frances, and this is Dwight. We are from California," she launched forth into a synopsis of her life. She was the daughter of a Rabbi, she spoke six languages, many of her people had been slain in Germany, etc. But the main thing to her was that she knew and loved Jesus! She also knew about the Convention and wanted to attend, but supposed she could not, without an invitation. So I issued one right then and there, and told her that she could sit with us and that I would help her register. All this time we were walking rapidly across the street and up the steps and into the large convention building. It's most symbolic to me that a daughter of Jerusalem whose name was Miriam, was leading me into the Convocation. And I, in turn, was escorting her. After all another daughter of Jerusalem, named Mary, has led me into holy places. And I too bear her with me.

The Binyanei Hauma was teeming with people and vitality. I doubt that it will ever be the same since that gathering! And it was indeed thrilling to be in the midst of so large a crowd of Pentecostal Christians. We are alone so much of the time, or in small groups; but whenever we are privileged to mingle with a Christian multitude, it thrills and uplifts us, and helps us to picture the great universal throng. The various nationalities were seated in sections. And the United States had such a huge delegation that they occupied most of the central part of the first floor. California led the parade—and had a whole section to itself—and to this we were directed. The seats were almost all taken so we had to sit far back. Directly behind us was a French speaking group from many parts of the world. What joy it was to walk in and hear their interpreter talking quietly to them in French! The German group was close by, and the Spanish, and the various other language groups were thus segregated. Their flags and banners were delightful to see. Of course most of the Africans and Orientals spoke English or French, so they were included in these sections.

Sweden had sent a very large group indeed, so it was the second language of the Convention. As soon as the meeting opened, we found that everything was said first in English, then in Swedish, and then simultaneously in all the other languages by their interpreters. You can imagine what a babel this made! It was a little like Pentecost, I imagine, with many tongues being heard at once. I enjoyed hearing the French translation especially, and was glad I was nearest to it. There was a short time of prayer, singing and music. And since Miriam did not know the songs, I explained to her what they meant. Meanwhile she was babbling constantly in my ear—so eager was she to tell me how much she loved me, how glad she was I had come, and how eager she was to be healed and to receive the baptism of the Holy Spirit. She quickly wearied of the inevitable introductions, acknowledgements and announcements that accompany such conventions. And she was very fearful about registering—thinking they would put her out. As soon as this came up I gave her the two dollars and helped her fill out the form which was circulated. When she received an identification card and the program, map and other literature that was given to each, she relaxed somewhat and began to praise the Lord. She could not see why ALL were not praising the Lord. And her concern about it began to annoy those sitting around us, who were trying to hear the speakers. Dwight was most uneasy, and finally I had to shush her gently. But I too was heavy of heart because there was no season of praise, and very little spirit of prayer manifested.

The meeting seemed long and tiresome to me, I am sorry to say. And dear Miriam was almost in tears because they talked so much and did not preach the Word or call the people to prayer. I tried to explain about the necessity for all these opening arrangements. But I was relieved when we were finally dismissed and told to return to our buses—so we could be taken to our hotels for lunch. We had caught a glimpse of Dr. McKoy, sitting way down in front, but in that crowd there was no way to reach him.

Lunch was very good indeed. But I was wrestling with a problem and getting heavy of heart. I had spoken to Bondi, our guide, about the Synagogue tour. And to my surprise he had seemed almost offended! And then he had remarked that he did not know where or when it was, and he thought I had been misinformed. When I told him that the Tour Office man had confirmed this information that very day, he simply shrugged. I felt that he was opposed to my going. So I dropped the subject. As soon as lunch was over I hurried to the hotel desk for help. But again I was rebuffed. The clerk was out and the secretary talked with me. She was a snobbish type, and she said she knew nothing at all about any such thing. I had sensed that she resented us, and I saw that she had no intention of helping me. (I guess everything we do in the Lord has to be tested.) I was near to tears when we went up to our room. And had I said a word to Dwight, I am sure he would have urged me to go with him. After all, we had been told that the Mayor of Jerusalem would welcome us—in the enforced absence of Premier Ben Gurion—and it would be a most important meeting.

I had to prepare by faith and appear cheerful until he went down to wait for the bus. Then I flew to getting ready. I felt a most wonderful sense of anticipation and of dressing for the Lord. So I wore my orchid suit and purple gloves and stole. (I was glad the weather had modified.) And I was careful to place my Star of David pin in a prominent place. Perfumed and dressed up—as for the King—I descended to the lobby, trying to appear poised and assured. But I was really trembling inside. To my dismay, I saw that the secretary was still at the desk. So I walked over to her and casually remarked, "Would you call me a taxi, *b'va'kashah* (please)? I had already learned that she did NOT like my using Hebrew words. She looked aghast! "Your bus has already gone," she said. "Yes, I know," I replied, "what I want is a taxi." So of course there was nothing to do but give me one. As I waited on the porch, I will admit that I was excited to get away from everyone, Dwight included, and venture forth alone with the Lord—a thing I have always delighted in. I had no idea what was going to happen, but I felt much expectation. When the cab arrived I told the driver to please take me to the Synagogue tour place. He laughed and said, "Well, it's only a short walk, but I will take you." I told him then how I had inquired about it and with what results. And he seemed dumbfounded. "Everyone in Jerusalem knows about the Synagogue tour," he said. "It is impossible that they should not know." My mouth was opened to witness to him about the Lord, and how I felt He wanted me to visit the synagogues, and he seemed most impressed and warm toward me. When we parted I said, "*Shabbat shalom,*" as the headwaiter had told me to do for the Sabbath, and I added, "*Oov'rachah*" (and a blessing!) And he blessed me too.

The lecture was held in what appeared to be a large school building near a Synagogue. I bought my ticket and went all the way to the front of a large hall. It was filling fast, and I felt impressed to be right in the heart of it. The seats around me soon were occupied by what appeared to be Jews from the United States—tourists too. And one of them immediately started talking with me. She said that she was sure I was one of the Pentecostal Convention women. And this amazed me. My little pin didn't fool her a bit. But she was most interested in why I was in the city and attending this lecture. And soon I was really pouring forth words of testimony, love and Scripture. Her husband, and others with them listened too. And more questions were asked. They could not understand why we celebrate Pentecost, for one thing, since this, they think, is a *Jewish* holy day! The Lord gave me love and wisdom in answering. And I was aware that the people in the row behind were listening too— and that I was practically preaching. About this time the Rabbi entered, and he was quite a character! They all agreed that he was worth crossing continents to hear— mainly because he was so "entertaining!" He proved to be a born comedian. Yet he was most devoted to the Lord in his typical Yiddish way. And soon I was indeed fascinated by his lecture, which I was told is never given the same way twice.

The Rabbi began by welcoming us and telling us about the wonderful privilege of being in Jerusalem for Pentecost. He mentioned its Old Testament meaning, and also told us about the traditional belief that David was born and died at this date. I tried at first to take notes, but he talked so fast, and so moved about the whole time— acting out everything he said—that I could not take my eyes off him without missing something. I could understand more about our demonstrations in the Spirit, after watching him! I was indeed thankful for our wonderful background in the Feasts and types displayed in Israel. I daresay that I knew more about what he was expounding than many of those American Jews. The Spirit has taught us most faithfully! Our Rabbi told us too about the different racial groups who had returned to Jerusalem, and how each worships with the same scriptural words, but to their own musical setting. He danced, pranced and demonstrated some more, singing in Hebrew, Yiddish and several other languages. It was most moving, and I was often near tears; yet at the next moment laughing heartily with the entire crowd. It was delightful indeed.

The thing that impressed me most, however, was what he said about the Sabbath. When sunset on Friday afternoon drew near—according to him—a miraculous change began to take place in every devout Jewish household. The father who worked hard all week, and may be grumpy at times, bathes and dons his best clothing and begins to radiate joy and praise, for the Sabbath is to be welcomed like a "bride" and embraced with rejoicing. This is the one day of the week that he is not only a father, but an actual king! The head of each house thus depicts the Lord Himself, as He shall be King of all the earth when this dispensation of work and trouble ends. All cares are to be forgotten. All thoughts of poverty are to be banished. He is rich, he is joyful, he is free, he is king! The mother too lays aside her apron and puts on her finest dress. She also adorns herself with jewels, for she, of course, is queen—a true queen in God. It is her joy to place spotless linen and the nicest dishes on the table for the Sabbath supper. Flowers are arranged, and the best foods the family can afford are already prepared and waiting for their enjoyment. It is she who lights the candles and gathers the family. It is a time for prayer, praise, joy and sweet wine. And until the next sunset the entire family is kept clean, well dressed and feeling like little princes and princesses.

As this was all described and depicted—in a most humorous and yet inspiring manner—I could not help but think of our own dear little Kingdom Company. We too have known these times of forgetting all labor and trials, donning our best garments, adorning ourselves for the Lord and then participating in foretastes of Kingdom splendor. We have entered into that rest—that Sabbath—prepared for the people of God. How flat and empty and commonplace the Christian "Sabbath" seems by comparison—among those of the formal church!

When the lecture ended, we were divided into several busloads and taken to the old *Mea She'arim* section of the city. There are many Synagogues here, all fairly close together. Our particular group was taken first to the Spanish Synagogue—a very old and ornate, though small, building. On our way to it, we had passed by the doorways and little courtyards of these devout Orthodox Jews. Children were standing about quietly, near most of these homes, clean and shining of face. Already they had been prepared for the Sabbath—now only about an hour away! The Rabbi had asked us not to give money to the children. "We don't want them to be beggars," he said. But we could give them gum or candy. How I wish I had some at hand! I could not help but contrast each Israeli child with the gay, dirty, friendly little "beggars" in Jordan. They had a way of radiating love even when not given a cent. And one of the most poignant memories I cherish of the entire trip was of a small, dirty Arab girl who suddenly ran after me—on the streets of Old Jerusalem—and tried to take my hand. I thought at first she was imploring me for a gift. But when I opened my hand, she placed in it about a dozen small nuts! Then, with the dearest smile and a "halo" she ran away; while I, wanting to cry, tried to call my thanks after her. I put those nuts away and still cherish them! There was none of this kind of eagerness and love manifested by the children we saw in the New Jerusalem—even though in this district they were poor, and their homes were very much like the ones on the Jordan side, except that these were clean. And so were the children. But also proud, and aloof, though polite. They answered our calls of *"Shabbat shalom!"* And some of them bowed. When I turned to wave and say, *"Oov'rachah"* they appeared embarrassed. But I blessed them anyway.

After seeing the Spanish Synagogue, we were taken to a very old Persian one. And this was interesting, I guess, though I was enjoying seeing the people and hearing the Hebrew conversation more than what each Rabbi explained as we entered. There were a few old men reading the Torah in both buildings. Our next visit was to a Yemenite Synagogue. It was very small, and yet most worshipful, I felt. And as soon as I entered I felt the manifest presence of the Lord. In one corner a very aged Yemenite brother was sitting cross-legged on the floor, teaching the Torah to a few children, much as Jesus must have been taught in His youth. They were all wearing Yemenite clothing. There was another Yemenite who appeared to me like a saint or prophet. (I believe we were told that he was in his nineties.) He too sat on the floor, eyes uplifted, worshiping the Lord. On his face was a supernatural glory—and what a face it was! In him I could see Moses, or any one of the ancient prophets. His eyes were like great pools of light—such eyes I have never seen before. He appeared pure and childlike, yet mature in wisdom and understanding. Perhaps the Lord put a special glory on him that day—it seemed to me like a theophany.

In the last synagogue, the fourth, we were given a special message by the Rabbi, who stood above us in a sort of enclosure. To my surprise, he began to speak of the Messiah. And he spoke of the hope of His soon coming. He said that He would ride into Jerusalem on a white ass or donkey, and that He would bring peace and deliverance to Israel. I can't tell you how odd it seemed to be hearing these words—just as though Jesus had not done this very thing! Then he prayed over us and pronounced a special blessing upon us. And we in turn wished him Sabbath peace—in Hebrew of course. Our tour now was ending, and I was led to tarry and be one of the last to leave. Outside I found that several of the women who had talked with me in the hall and on the bus enroute were waiting for me. One woman, who had come there for the wedding of her son, asked me a pointed question. What did I believe about the Messiah? She said that her son was very religious and that he worried about this question. The others crowded around me to hear the answer, and the Lord truly spoke forth and again drew me. I cannot remember much of what was said. But I do know that I testified of Him, and that He had indeed come in just the manner the Rabbi had described, and how He had died for them and for us all. She didn't seem troubled and said, "What will He do with us, if this is so?" The others too murmured about how they had been punished so much, and wondered if rejecting Jesus could be the cause. I assured them that He loved them now just as He had always loved them, and that He would return again. I cannot seem to stop talking about Him, nor did I try. And they stood and listened until the guide called back for us to hurry on—it was almost sunset, and we must be out of the quarter before Sabbath began. It was then indescribable joy to stand tough on the streets of New Jerusalem and "prophesy" of the Messiah to these women, even as I had walked the streets and prophesied of Him in Old Jerusalem, when the Spirit transported me there years ago.

A hush came over all of us as we sensed the mysterious coming of the "the bride," the Sabbath, to Mea She'arim, and to all Jerusalem. The woman who had questioned me slipped her arm around me, and we walked over the worn cobbled pavement in silence. And then a very dramatic thing happened! Suddenly, a young man, garbed in a biblical sort of costume, appeared around a corner. And in his hand was a shofar, a ram's horn. I almost held my breath as we paused to watch him lift it to his lips and blow several long, clear blasts. A rustle ran through our little band of women. "It is time for the Sabbath to begin," they whispered to me, "and we must hurry away." Then I understood why the young man had scowled at us. But my heart was full of laughter. This was a rare and most precious moment—one the Lord Himself had arranged. The Spirit had trumpeted forth through my voice. And then the ancient shofar had been sounded in the holy sector of the Holy City. And the trumpet was sounding in my soul! I knew the Lord had been faithful to keep His appointment with me on this wonderful day which He Himself made for us to be glad and rejoice about.

You will all want to know about how the day ended, of course. And even though it is somewhat anticlimactic, it was nonetheless most rewarding. The bus delivered each visitor to his own hotel. I got off in a few blocks, and I recalled that the dear lady who had questioned me and walked with me, was looking back and waving at me as long as she could see me. I felt a sweet assurance for her salvation. And of course I looked to the Lord for all of them who had heard, including her son who would soon be hearing through her.

Dwight had already returned to the hotel, and he was so full of what had happened in the afternoon meeting that I could say but little about my own wonderful experiences. I felt too that—as with many things the Lord does—He would be pleased if I cherished those things and pondered them in my heart, and later shared them with you who went with me in Spirit. I noticed too that my voice was very hoarse again. And then I realized that all afternoon I had spoken in a clear voice, with no trace of the hoarseness of my cold!

As for Dwight—well, the Mayor had spoken to them at length, and so had other dignitaries. They had given to each delegate a beautiful bronze medallion—with an ancient seal engraved on it, and the words, "Out of Zion shall go forth the Law, and the Word of the Lord from Jerusalem"—a verse very familiar to all of us! On its reverse side these words also were engraved: "Presented by the Government of Israel to the members of the Sixth Pentecostal World Conference, Jerusalem, Israel, May 19-21, 1961." I was thrilled about this too—for on the top was a Candlestick much like ours. And I was indeed glad that Dwight had been able to get one for me, even though I was absent.

After a short rest period, in which I enjoyed your blessed letters, which had begun to catch up with me, I did some more laundry and cleaned up for special dinner with Dr. McKoy. The King David was only a few short blocks from our hotel. But Dr. McKoy had insisted on calling for us and taking us by taxi, if I remember correctly. Anyway, we arrived in a very happy mood. And our apostle seemed almost beside himself with joy. He explained that not very often could he entertain, but that once in a while the Lord led him to do something like this, and show kindness and love to his dear ones. And I truly felt that it was the Lord who led us into the elegant lobby of this famous hotel, and on into its spacious dining room. Nothing less than filet mignon was in order, he told us, and we were quite willing to agree. Steaks are very choice in Israel—and in all that part of the world. This was royal fare. With Dr. McKoy was his dear friend and "son" in the Lord, Emil Olson. He had been going to school and ministering in personal work in Israel for several months. So there was much of interest to talk about. But I sensed that Dr. McKoy heard very little. And this was painful to me, for I was so hoarse again by now that I had to almost croak to be heard at all. So he and I settled back and enjoyed the dinner, leaving Dwight and Emil to do all the talking.

I was very happy indeed to do this for, after all, the Lord and I had already had our talking time that afternoon. And now I could just rest, eat and absorb His wonderful love which was flowing from all our hearts. After we finished this really lavish feast, we strolled into the lobby and sat and visited in the regal looking chairs—quite as though we were really being entertained by the King, which we were! When the time came to part, Dr. McKoy had persuaded Dwight that we just must go with him to the Synagogue for Sabbath service on the following morning. So we agreed to join him right after breakfast. Then we walked slowly back toward the hotel as though in a dream. This had been one of life's few perfect days—a day spent not only in time, but in eternity.